Lecture Notes in Computer Science 5176

Commenced Publication in 1973
Founding and Former Series Editors:
Gerhard Goos, Juris Hartmanis, and Jan van Leeuwen

T0223184

Sven Hartmann   Xiaofang Zhou
Markus Kirchberg (Eds.)

Advances in
Web Information
Systems Engineering

WISE 2008 International Workshops
Auckland, New Zealand, September 1–4, 2008
Proceedings

 Springer

Volume Editors

Sven Hartmann
Clausthal University of Technology
Julius-Albert-Str. 4
38678 Clausthal-Zellerfeld
Germany
E-mail: sven.hartmann@tu-clausthal.de

Xiaofang Zhou
The University of Queensland
Brisbane QLD 4072
Australia
E-mail: zxf@itee.uq.edu.au

Markus Kirchberg
Institute for Infocomm Research A*STAR
21 Heng Mui Keng Terrace
119613 Singapore
E-mail: mkirchberg@i2r.a-star.edu.sg

Library of Congress Control Number: Applied for

CR Subject Classification (1998): H.4, H.3, H.2, C.2.4, I.2

LNCS Sublibrary: SL 3 – Information Systems and Application,
incl. Internet/Web and HCI

ISSN   0302-9743
ISBN-10   3-540-85199-2 Springer Berlin Heidelberg New York
ISBN-13   978-3-540-85199-8 Springer Berlin Heidelberg New York

Springer is a part of Springer Science+Business Media

springer.com

© Springer-Verlag Berlin Heidelberg 2008
Printed in Germany

Typesetting: Camera-ready by author, data conversion by Scientific Publishing Services, Chennai, India
Printed on acid-free paper   SPIN: 12459407   06/3180   5 4 3 2 1 0

Preface

The International Conference on Web Information Systems Engineering (WISE) provides an annual forum for exploring research, development, novel applications and industrial innovations in the area of Web Information Systems. The 9th edition of this conference (WISE 2008) was held in Auckland, New Zealand from September 1 to 3, 2008. This volume contains the papers that were presented during the WISE 2008 workshops. We commend these papers to you and hope you find them useful.

A major objective of the WISE conferences is to identify new issues and directions in Web engineering, to share experiences, to host discussions, and to initiate future work and collaborations. Associated workshops devoted to emerging or specialist topics are an important part of the WISE conferences helping to make them an inspiring experience for all participants. Three workshops were organized and held in conjunction with the WISE 2008 main conference:

- The First International Workshop on Web Information Systems Engineering for Electronic Businesses and Governments (E-BAG 2008), chaired by Sebastian Link (Victoria University of Wellington, New Zealand), Hui Ma (Victoria University of Wellington, New Zealand), and Jian Yang (Macquarie University, Sydney, Australia);
- The Second International Workshop on Web Usability and Accessibility (IWWUA 2008), chaired by Silvia Abrahão (Valencia University of Technology, Spain), Cristina Cachero (University of Alicante, Spain), and Maristella Matera (Politecnico di Milano, Italy); and
- The First International Workshop on Mashups, Enterprise Mashups and Lightweight Composition on the Web (MEM&LCW 2008), chaired by Marek Kowalkiewicz (SAP Research Brisbane, Australia), Dominik Flejter (Poznan University of Economics, Poland), and Tomasz Kaczmarek (Poznan University of Economics, Poland).

Following calls for papers, we received 40 submissions. All submitted papers were carefully reviewed by at least three international experts, and the best 16 were then selected for presentation during the workshops and for inclusion in this volume. We would like to thank all authors of submissions, and the members of the program committees of the WISE workshops for their timely expertise. We are grateful to the workshop organizers for the time and effort they spent to guarantee the high quality of the program, thus contributing to the success of WISE 2008. We further thank Athman Bouguettaya (CSIRO, Australia) and Giorgio Brajnik (University of Udine, Italy), who kindly agreed to give keynote talks to the workshop participants.

High-quality tutorials are a long-standing tradition of the WISE conferences. They complement the depth-oriented regular paper sessions and keynote talks by providing conference participants with broad overviews of emerging fields or clear syntheses of results in existing fields. For WISE 2008, we invited Hans-Ludwig Hausen (Fraunhofer National Research Center, Germany), and Shazia Sadiq and Ke Deng (The University of Queensland, Australia) to present tutorials on specific areas of their expertise in Web technologies and methodologies. We wish to express our appreciation to the tutorial speakers for their illuminating presentations and their patience in answering questions from the interested audience.

Special thanks are due to Yanchun Zhang (Victoria University of Melbourne, Australia) for moderating the WISE 2008 panel discussion on "Engineering Issues for the Web 2.0".

We would like to thank all those who helped to make WISE 2008 a remarkable event. We are particularly grateful to the local organizers at Auckland University of Technology for the wonderful days in New Zealand.

September 2008   Sven Hartmann
                                                                   Xiaofang Zhou
                                                                   Markus Kirchberg

WISE 2008 Conference Organization

General Conference Co-chairs

Schewe, Klaus-Dieter   Information Science Research Centre, New Zealand
Wang, Xiaoyang Sean   University of Vermont, Burlington, USA

Program Committee Co-chairs

Bailey, James   The University of Melbourne, Australia
Maier, David   Portland State University, Portland, USA
Thalheim, Bernhard   Christian-Albrechts-University Kiel, Germany

Workshop and Tutorial Co-chairs

Hartmann, Sven   Clausthal University of Technology, Germany
Zhou, Xiaofang   The University of Queensland, Brisbane, Australia

Local Organization Co-chairs

Parry, Dave   Auckland University of Technology, New Zealand
Pears, Russel   Auckland University of Technology, New Zealand

Panel Chair

Rossi, Gustavo   Universidad Nacional de La Plata, Argentina

Industrial Program Co-chairs

Hosking, John   University of Auckland, New Zealand
Pastor, Oscar   Universidad Politécnica de Valencia, Spain

Steering Committee Liaison

Zhang, Yanchun   Victoria University, Australia

Publicity Chair

Kirchberg, Markus   Institute for Infocomm Research, A*STAR, Singapore

Organized by

WISE Society
c/o Department of Computer Science
City University of Hong Kong
Tat Chee Avenue, Kowloon
Hong Kong, China

http://www.wisesociety.org/

Auckland University of Technology
New Zealand

http://www.aut.ac.nz/

**Information Science Research
Centre**
20a Manapouri Crescent
Palmerston North 4410
New Zealand

http://isrc.mucoms.org/

Supported by

**Institute for Infocomm Research
(I^2R)**
Agency for Science, Technology and
    Research
21 Heng Mui Keng Terrace
Singapore 119613
Singapore

http://www.i2r.a-star.edu.sg/

**Christian-Albrechts-University
Kiel**
Germany

http://www.uni-kiel.de/

WISE 2008 Workshop Organization

WISE 2008 Workshop and Tutorial Co-chairs

Hartmann, Sven   Clausthal University of Technology, Germany
Zhou, Xiaofang   The University of Queensland, Brisbane, Australia

E-BAG 2008 – First International Workshop on Web Information Systems Engineering for Electronic Businesses and Governments

Workshop Chairs

Link, Sebastian   Victoria University of Wellington, New Zealand
Ma, Hui   Victoria University of Wellington, New Zealand
Yang, Jian   Macquarie University, Sydney, Australia

Publicity Chair

Kirchberg, Markus   Institute for Infocomm Research, A*STAR, Singapore

Program Committee

Abramowicz, Witold   The Poznan University of Economics, Poland
Chiu, Kak Wah   The Chinese University of Hong Kong, China
García González, Roberto   University of Lleida, Spain
Griffiths, Mary   The University of Adelaide, Australia
Gritzalis, Stefanos   University of the Aegean, Greece
Hartmann, Sven   Clausthal University of Technology, Germany
Huang, Joshua Zhexue   The University of Hong Kong, China
Janssen, Marijn   Delft University of Technology, The Netherlands
Katasonov, Artem   University of Jyväskylä, Finland
Kirchberg, Markus   Institute for Infocomm Research, A*STAR, Singapore
Li, Xue   The University of Queensland, Australia
Lu, Jianguo   University of Windsor, Canada
Park, Sang Chan   KAIST, Korea
Parry, David   Auckland University of Technology, New Zealand
Polemi, Despina   University of Pireaus, Greece
Stojanovic, Ljiljana   Research Centre for Information Technologies Karlsruhe, Germany

| Stormer, Henrik | University of Fribourg, Switzerland |
| Wang, Jiying | City University of Hong Kong, China |

External Referees

Starzecka, Monika
Tao, Aries

IWWUA 2008 – Second International Workshop on Web Usability and Accessibility

Workshop Chairs

Abrahão, Silvia	Valencia University of Technology, Spain
Cachero, Cristina	University of Alicante, Spain
Matera, Maristella	Politecnico di Milano, Italy

Program Committee

Abou-Zahra, Shadi	World Wide Web Consortium (W3C), France
Bevan, Nigel	Professional Usability Services, United Kingdom
Bolchini, Davide	University of Lugano, Switzerland
Calero, Coral	University of Castilla-la-Mancha, Spain
Casteleyn, Sven	Vrije Universiteit Brussel, Belgium
Catarci, Tiziana	Università degli Studi di Roma 'La Sapienza', Italy
Cavalcanti Leite, Jair	Universidade Federal do Rio Grande do Norte, Brazil
Centeno, Vicente Luque	University Carlos III, Spain
Daniel, Florian	Politecnico di Milano, Italy
Gaedke, Martin	Chemnitz University of Technology, Germany
Gaffney, Gerry	Information & Design, Australia
Garzotto, Franca	Politecnico di Milano, Italy
Houben, Geert-Jan	Vrije Universiteit Brussel, Belgium
Insfran, Emilio	Universidad Politécnica de Valencia, Spain
Law, Effie Lai-Chong	ETH Zürich, Switzerland
Lozano, Maria Dolores	University of Castilla-la-Mancha, Spain
Mendes, Emilia	University of Auckland, New Zealand
Olsina, Luis	Universidad Nacional de La Pampa, Argentina
Poels, Geert	University of Ghent, Belgium
Santoro, Carmen	ISTI-CNR, Italy
Trujillo, Juan Carlos	University of Alicante, Spain
Vanderdonckt, Jean	Université catolique de Louvain, Belgium
Winckler, Marco	University Paul Sabatier, France

External Referees

Heil, Andreas
Moraga, Mª Ángeles

MEM&LCW 2008 – First International Workshop on Mashups, Enterprise Mashups and Lightweight Composition on the Web

Workshop Chairs

Kowalkiewicz, Marek	SAP Research Brisbane, Australia
Flejter, Dominik	Poznan University of Economics, Poland
Kaczmarek, Tomasz	Poznan University of Economics, Poland

Program Committee

Brickley, Dan	University of Bristol, United Kingdom
Caceres, Marcos	Queensland University of Technology, Australia
Dreiling, Alexander	SAP Research, Australia
Ferraiolo, Jon	IBM & OpenAjax Alliance, USA
Grassel, Guido	Nokia Research Center, Finland
Jatowt, Adam	Kyoto University, Japan
Rodger, Richard	TSSG, Italy
Zheng, Yu	Microsoft Research Asia, China

Table of Contents

Regular Papers

MEM&LCW 2008 – First International Workshop on Mashups, Enterprise Mashups and Lightweight Composition on the Web

Regular Papers

WISE 2008 Panel

E-BAG 2008 Workshop PC Chairs' Message

Sebastian Link[1], Hui Ma[1], and Jian Yang[2]

[1] Victoria University of Wellington, New Zealand
[2] Macquarie University, Sydney, Australia

Web-based information systems offer tremendous opportunities for both business and governments. The Web has provided new possibilities for companies to communicate and engage in business, and for governments to exchange information and services with citizens and organisations. On one hand, the implementation of individual objectives in E-Commerce and E-Government demand new Web technologies and original methodologies for engineering quality information systems. On the other hand, technological and methodological advances can reveal new openings to conduct electronic business or improve the efficiency, convenience, and accessibility of public services.

This volume contains the papers presented at the First International Workshop on Web Information Systems Engineering for Electronic Businesses and Governments (E-BAG 2008) which was held in Auckland, New Zealand from September 1 to 3, 2008 in conjunction with the Ninth International Conference on Web Information Systems Engineering (WISE 2008).

The main purpose of the workshop is to assess current approaches, techniques and practices by which web information systems implement the objectives of electronic businesses and governments. We aim at bringing together researchers and practitioners that are interested in exchanging experiences and ideas about utilising/engineering Web information systems for the purpose of E-Business and E-Government. The scope of E-BAG 2008 included topics such as:

- Principles and Foundations;
- Languages and Models;
- Technologies;
- Challenges and Issues;
- Management and Strategy;
- Ontologies;
- Semantic Web;
- Agents;
- Collaboration;
- Retrieval and Search;
- Integration and Mediation;
- Mining and Discovery;
- Privacy, Security and Trust;
- Social, Cultural and Consumer Aspects;
- Training and Education;

S. Hartmann et al. (Eds.): WISE 2008, LNCS 5176, pp. 1–2, 2008.

- Mobility and Ubiquitousness;
- Preservation and Quality;
- Law and Ethics; and
- Case Studies.

Following the call for papers we received 9 submissions. There was a rigorous refereeing process that saw each paper refereed by at least three international experts. The best three papers, as judged by the program committee, were accepted and are included in this volume. Moreover, we added two papers that were recommended to us from the WISE 2008 conference.

We are grateful to Dr Athman Bouguettaya from the Australian Commonwealth Scientific and Research Organization ICT Centre at Canberra, Australia, who presented the E-BAG 2008 keynote address on *Service Computing for the Service Economy*.

We wish to thank all authors who submitted papers and all workshop participants for the fruitful discussions. We also like to thank the members of the program committee for their timely expertise in carefully reviewing the submissions, and Markus Kirchberg for his excellent work as E-BAG publicity chair. Finally, we wish to express our appreciation to the local organisers at the Auckland University of Technology for the wonderful days in New Zealand.

Service Computing for the Service Economy

Athman Bouguettaya

CSIRO
ICT Center
Canberra, Australia
athman.bouguettaya@csiro.au

World economies have effectively moved to a service economy where at least 75% of the GDP of most western countries is in the service sector. A large percentage of the service economy is in the government and commerce sectors. For instance, advances in e-government and e-commerce technologies have opened up new markets to provide client-centric services which usually require, in many instances, new and novel models of collaborations for service providers.

It is noteworthy to mention that every major shift in the economy has traditionally be accompanied and/or stimulated by new technologies that support that shift. Although the service economy has been very pronounced for at last a decade, there have been very little focused research efforts to support and enhance it. This realization has inspired a flurry of research activities both in industry and academia to fill the void. As a result, there has been a strong push to build a foundation for a new science, called *service science* to cater for this new economic paradigm. A major supporting technology is the emerging service computing that would represent services as computing artifacts using Web services, thanks largely to the near industry consensus around the key standards for specifying Web services (WSDL), registering and advertising Web services (UDDI), and modes of communications (SOAP).

Web services are expected to be the key technology in enabling the next instalment of the Web in the form of the Service Web. In this paradigm shift, Web services would be treated as first-class objects that can be manipulated much like data is manipulated in a database management system. While initial standards have been beneficial in the early adoption and deployment of Web services, innovations and wider acceptance of Web services need a rigorous foundation upon which applications and systems can be build. There is a strong impetus for defining a solid and integrated foundation that would stimulate the kind of innovations witnessed in other fields, such as databases. Materializing this vision requires solutions to the different fundamental research problems to deploying Web services that would be managed by an integrated Web Service Management System (WSMS).

Fully delivering on the potential of next-generation Web services requires building a foundation that would provide a sound design for efficiently developing, deploying, publishing, discovering, composing, and optimizing access to Web services. The proposed Web service foundation will enable the development of a uniform framework that would be to Web services what DBMSs have been to

S. Hartmann et al. (Eds.): WISE 2008, LNCS 5176, pp. 3–4, 2008.

data. In this framework, Web services will be treated as first-class objects which can be manipulated as if they were pieces of data. In this talk, I will first give an overview about services and the need for a service science. I will then motivate the need for treating services as a first class objects. I will then overview our own research work developing the foundation of the core components of WSMSs which include: Web service query optimization, Web service composition, Web service change management, and Web service trust management. Finally, I will overview an E-government WSMS prototype that has been used as a deployment test-bed.

Minimizing the Impact of Change on User Productivity

Ratvinder Singh Grewal[1], Barbara Targonski[2], and Quoc Hao Mach[1]

[1] E-Business Science, Laurentian University, Sudbury,
Ontario, Canada
{rgrewal,qh_mach}@laurentian.ca
[2] Sandvik Mining and Construction Global IT, 100 Magill Street,
Lively, Ontario, Canada
barb.targonski@sandvik.com

Abstract. In order to stay abreast of technological advancements in the business world, companies must weigh the benefits of adopting new technology with the setbacks associated with new implementation and training. Understanding the interactions between the end users and new technology is crucial for a successful transition from the present day technology to future technology. Users are required to learn new technology quickly, retaining knowledge of old technology while transferring this knowledge base to the new technology. Understanding of good design and interpersonal user characteristics as well as use of mental models can have a significant impact on maintaining user expert level when the change is introduced. Ability to maintain high expertise among users and use knowledge transfer will ultimately result in diminishing the time it takes to reach an expert level and translate to minimizing costs related to learning.

Keywords: Technology adoption, change management.

1 Introduction

The rapid change of technology and its ubiquitous integration into everyday life has intensely widened the range of users that need to be considered in technological design. Systems and technologies already in place must be able to adapt quickly and reliably to take advantage of new business opportunities to remain competitive. Users who have gained expert status with current technology must be able to quickly adapt and learn the new system when new technology is introduced. While change is inevitable, the way it is executed and the approach taken must respect user characteristics and mental models. These play a key role in successful technology change introduction.

With the business environment in constant flux, the integration of new technology is critical. Systems that operate in the real world must be able to adapt continuously in order to satisfy business, user and customer demands. Technological change can occur by adapting a system that is already in place or upgrading the current systems. The change that is adopted will be driven by the business requirements and the development of new technology. The reasons for changes may originate from legacy upgrades, standardization, new version release, service agreement replacement or adopting a more suitable solution. When these changes do occur, the benefits of the

S. Hartmann et al. (Eds.): WISE 2008, LNCS 5176, pp. 5–11, 2008.

new technology have to be weighed against the learning of the new equipment. Researching the human component of the equation will assist us in determining whether the new technology can be easily integrated with the user's previous knowledge, or determine whether a new skill set must be learned. This can assist a business in determining whether the new technology should be adopted.

2 Technology Adoption

In the past, users were forced to adapt to the technology. Systems were designed with the programmer or designer in mind and not the end user, leading to usability difficulties. The new thought process now focuses on increasing usability for the end user when developing new technology. With the advancement and availability of computers, the demographics of the users have changed. Computer users are no longer required to be technically advanced and highly educated within the computer field. This reliance on computers has grown, with computerized components playing integral roles in phones, cars, airplanes, supermarkets and almost every facet of everyday living. The result of this inclusion forces the designers to produce technology that is usable for a wide range of users who may be less technical, more demanding, explorative, and impatient.

When a new interface or device is introduced, a natural progression in the skill level development curve from a novice to an expert level can be observed. This progress can be referred to as learning curve which reflects the relationship between the duration of learning or experience and the resulting progress. When users are first introduced to software or a device with no previous knowledge they are considered a novice. Because of human factors such as intelligence and personal traits, users will progress on the learning curve at different rates. User's progress on the learning curve slope upwards as the time advances, until they reach a peak or plateau, where learning stops, this can be classified as expert status. This is when the user reaches an automated like response or the subconscious thought pattern become a series of actions that proceed rapidly and automatically without effort [1]. Once the maximum learning potential is reached, one of two things can happen: the individual will maintain the expertise level or the effective usage of system will decline either due to boredom or introduction of change.

In business terms this learning curve can be referred to as a "cost improvement curve" or "efficiency curve" and this curve has a direct impact on the company's financial burden associated with change. The success of technology adoption is dependent upon users accepting the new technology. One of the models used to study whether new technological will be readily adopted by users is the technology acceptance model (TAM). This theory models how users come to accept and use the newly introduced technology. The model suggests that individual's behavioural intention to use the system is determined by perceived usefulness (PU) and perceived ease of use (PEOU). The TAM has been proven robust in predicting user acceptance of IT, and has been applied in understanding the motivational issues in computer and software adoption [2].

Factors can influence user acceptance either directly or indirectly through subjective norms, job relevance; the ability to demonstrate significant results on PU, computer self-efficacy, perceptions of external control, computer anxiety, and objective usability affecting PEOU [2]. One of the key factors influencing the adoption process is the ability to transfer knowledge from one application to the newer one [3]. By transferring existing knowledge from a familiar domain (the base) to a new product (the target), users can learn about a new product. Research in knowledge transfer and analogical reasoning suggests that this learning occurs through a series of stages: access, mapping, and transfer [3].

Some users have an intrinsic motivation to perform an activity for the sake of enjoyment of the activity. These users who possess this intrinsic motivation enjoy exploring and using the new features. Evidence has shown that this internal drive can play a large role in a user's adoption of new software [4]. This general understanding of the target user's characteristics will play an important role in determining whether the user will accept the new technology. The inclusion of devices facilitating tasks in business and normal daily activities has become pervasive. The multitude of variety and usage in different environments requires the user to be familiar with all of the devices. When new upgrades or technology are introduced in the business environment, consistency of application design must be considered. While inconsistency in the design can increase the time needed to learn a new application, the object must be designed and positioned in accordance with users' tasks. However, the work of the user who interacts with multiple applications or is recently introduced to an upgrade or replacement can be greatly simplified if the applications display a certain level of consistency. Interface consistency is an important goal that supports the search for interface properties that would lead to successful design.

When users learn a system, there is a progression through a learning curve as seen in Figure 1. During the first portion of the learning curve, users can be considered a novice user until they reach the maximum point of the curve where their learning plateaus and they obtain expert user status. If a business changes and introduces new technology or software the user must learn the process from a novice state with the progression through a new learning curve. If the user has obtained expert status and remains stuck at this state for an extended period of time, with the onset of boredom they may lose interest and their performance may degrade over time.

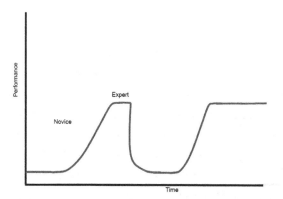

Fig. 1. Learning curve for Novice-Expert-Novice cycle

This is evident from Figure 2, where the user has reached a plateau and maintains this state for a while, followed by a decrease in performance over time. The way to prevent the cycle of novice-expert-novice is to alter the technology so transference of knowledge from system to the next system is possible. This will reduce the novice to expert to novice cycle, and lead to less downtime for the integration of the new system. The introduction of upgrades and changes to the system will also prevent boredom for the user, with the end result being higher levels of productivity over a longer time period.

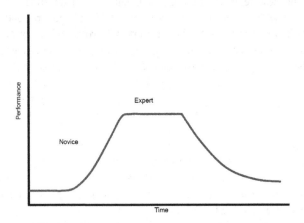

Fig. 2. Novice to Expert Cycle with performance decrease

Figure 3 illustrates the slight drop in performance when an expert user adjusts to the new system; however, the drop in performance is not as great when the change in technology involves some consistency in design to previous systems.

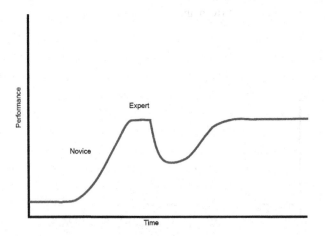

Fig. 3. Novice-Expert-Expert cycle with less performance decrease

Software engineering and technology development are primarily concerned with development of systems and devices that satisfy functional and non-functional requirements. Developers/designers are not only influenced by presented requirements and constraints but are highly aware of the mental models and social processes described earlier, embedded within organizational and cultural structures. The design process is not a purely technical task, therefore social and cognitive issues should be addressed during the design and technology selection process in order to provide users with the most beneficial learning and adoption environment.

Compared to designing and constructing buildings, cars, or washing machines, software development is still an immature discipline, and the way software is designed is constantly changing [5]. The goal of software development is to provide a tool that aids the user accomplish a goal or solve a problem in the most efficient manner. A designer's knowledge of human psychology and the knowledge of how things work are critical in this design. There is always a trade off between the beauty and the utility of the design. Non-functional requirements may contain visual requirements; however, if the utility of the design is compromised it will have an impact on the performance. Systems that can not be easily understood and adopted by users results in frustration and ultimately translates into collateral damage. Consistency in use of application may be considered one of the most important factors in user interface applications design.

Good conceptual design model plays a key role in creating systems that are easily understood by users and determines their usability. Conceptual design can be broken down into three components, the design model, the user's model and the systems model. The design model is the vision that the designer has in mind during the design phase. The user model is the conceptualization of how everything works to explain the functionality of the operation. Ideally the designer and the user's model is identical leading to an increased ease of use. The designer communicates to the user through the system, placing emphasis on the operation, appearance and functionality of the system. When designing the system, the designer must assure that the product has consistency and operates within the conceptual model created by the user. The best way to guide the user from novice to an expert status is to ensure that the systems model is less visible and a strong user's mental model of the system is used. Ensuring high correlations between the conceptual design model and the user's mental model leads to fewer errors and prevents the user from incorrectly interpreting how the system works. Patton (2007) suggests that in absence of a strong user's mental model, designers tend to self-substitute which may lead to personal bias in model creation and deviation from user-centric design. Software interfaces should be designed to help users build productive mental models of a system. Common design methods employed to support and influence users' mental models include: simplicity, familiarity, availability, flexibility, feedback, safety, and affordances.

Typically, the burden was placed on the user to learn how a software application works. In order to improve user satisfaction and system usability the burden shifted to the system designer. The system designers are now required to analyze and capture the user's expectations and build that into the system design [1]. Proper design should keep the user's goals and expectations in mind and aim to create simplified systems. Simpler systems not only increase the users' ability to learn and promote knowledge transfer but increase users' satisfaction.

New technologies such as personal computers are complex and an element of uncertainty exists in the minds of decision makers with respect to the successful adoption. People form attitudes and intentions toward trying to learn the new technology prior to initiating efforts directed at using the technology. Attitudes towards usage and intentions to use may be ill-formed or lacking in conviction or else may occur only after preliminary strivings to learn to use the technology have evolved. Thus, actual usage may not be a direct or immediate consequence of such attitudes and intentions [2].

The ability to change and adapt to the constantly changing business environment is critical and necessary to business longevity and competitiveness. Change is a difficult and expensive process. The cost of technology change is not solely associated with the capital investment regarding hardware and software but all the additional resources needed for successful implementation of change. Post implementation effects also include the adoption period of users to new technology. Human characteristics such as interpersonal skills, past experiences, social and cultural predispositions greatly impact the way users react to the new system. Understanding human expectations and the way knowledge is gained and transferred is critical when designing successful systems. Practicing principles of good design should always be considered when creating new technology but since the users are the central focus and ultimately the final approving factor, their expectation should be considered and involvement welcomed. Adoption of new technology by the users may be the ultimate test for any new design. Maximizing interface consistency may have a significant impact on decreasing the learning period required to adopt new application or upgrade. However consistency can not be perceived as a major focus in the application design, as functionality of the application is a driving factor in successful design. In order to minimize the costs associated with the learning period when new technology is introduced proper fit between the selected solution, the business needs and user characteristics is required. Gaining users' support and involvement in the implementation process as well as obtaining feedback that can be incorporated into the design itself will further contribute to greater acceptance and therefore easier learnability of the solution.

3   Conclusion

In order for successful integration of technology into a business, the most important aspect of change is the end user. In order to successfully merge the technology with the business the impact that change has on the user must be considered. This requires knowledge of both the learning process involved for the end user and the design process of the technology. The reduction of the learning in the adoption of new technology will lead to an increase in productivity. With both psychological and technological acumen, businesses adopting new technology will be a seamless transition.

References

1. Norman, D.A.: The Design Of Everyday Things. Basic Books, New York (1988)
2. Chan, H.C., Teo, H.: Evaluating the Boundary Conditions of the Technology Acceptance Model: An Exploratory Investigation. ACM Transactions on Computer-Human Interaction, Article 9 14(2) (2007)

3. Moreau, C.P., Lehmann, D.R., Markman, A.B.: Entrenched Knowledge Structures and Consumer Response to New Products. Journal of Marketing Research 38 (2001)
4. Chintakovid, T.: Factors Affecting End Users' Intrinsic Motivation to Use Software: 2007 IEEE Symposium on Visual Languages and Human-Centric Computing (2007)
5. Patton, J.: Understanding User Centricity. IEEE Software (November/ December 2007)

Mobile Payment: Towards a Customer-Centric Model

Krassie Petrova

Auckland University of Technology, 55 Wellesley Str. East,
Auckland, New Zealand
krassie.petrova@aut.ac.nz

Abstract. Mobile payment normally occurs as a wireless transaction of monetary value and includes the initiation, authorization and the realization of the payment. Such transactions are facilitated by purpose-built mobile payment systems that are part of the service infrastructure supporting the functioning of mobile business applications. A number of stakeholder groups may be involved in concluding a mobile payment transaction, among them customers, mobile operators, financial institutions, merchants, and intermediaries. In this paper, mobile payment systems are characterised from the point of view of the stakeholder groups. Building on existing work, a supply and demand model for the investigation of mPayment services is presented, and applied to a case study.

Keywords: Mobile payment, mPayment, mobile commerce, stakeholders, value chain, customer-centric, adoption.

1 Introduction

Mobile payment (mPayment) can be defined as a wireless transaction of monetary value which includes payment initiation, payment authorization, and payment realization. It occurs between a customer and a service or product seller (merchant). The transaction is carried via a mobile device connected to a mobile subscriber network such as a mobile phone. mPayment systems can be characterised using a number of defining features the most important of which are: the transaction amount, the payment settlement mechanism, and the mPayment supporting technology [1-4].

With respect to the transaction amount, an mPayment is either a macro-or micro-payment. A micro-mPayment is normally less than US$10.00, and is typically used to pay for mobile content (e.g. a mobile game). It is usually facilitated by a Mobile Network Operator (MNO) or a Mobile Subscription Service Provider (MSSP) through the billing mechanism. Macro-payments are larger and may need proper authorization by a bank or another financial institution; they are normally facilitated by a payment system set in place by the MNO, by a bank, or by a third party such as a Mobile Payment Solution Provider (MPSP) [1-2].

With respect to the payment settlement mechanism, mPayments can be classified as subscription account based (where the transaction amount is debited from or billed to the mobile subscriber' account), and card-based (the transaction amount is debited or billed to a credit/debit card) [4]. An example of a subscription-based mPayment is paying for a parking space at the point of parking. Online shopping using mobile

S. Hartmann et al. (Eds.): WISE 2008, LNCS 5176, pp. 12–23, 2008.

access to the Internet illustrates the second type of mPayment. Both mechanisms can be used for either a micro- or macro-payment.

Depending on the type of the supporting technology, mPayments are classified as contactless, and remote. 'Contactless' (or 'proximity') mPayments are conducted with the customer physically present at the point of sale and can be 'manned' (or 'face-to-face'), or 'unmanned' (or 'machine-to machine', for example buying refreshments from a vending machine). Proximity payments require an interface between the mobile phone and the merchant's payment terminal. Transactions carried remotely over the network (for example, downloading a news article) are referred to as 'over the air' (OTA). Conducting eCommerce over the mobile Internet and transferring funds are examples of OTA [5-7].

Even a brief overview of mPayment transaction types and settlement and technology characteristics shows that a significant number of market players can be involved in bringing an mPayment service to the customer. The number grows when mPayment is considered as an enabler of another mobile commerce (mCommerce) service. Third parties such as intermediaries may also get involved, for example to bundle a mobile service with a payment service or to provide customer authentication and payment authorization [2], [4], [8].

There is some evidence in the literautre to indicate that while mobile device penetration in New Zealand is suffciently high and on par with other developed countries, mPayment adoption and spread are still lagging behind (somewhat similar to Nordic countries in 2001-2002) [2], [8], [9]. The processes of mPayment acceptance and adoption have been studied widely and factors affecting consumer decisions to use mPayment and their managerial implications have been identified. However the dynamics of the process of meeting customer needs and preferences (demand) by the supply (the gammut of mPayment industry players) has not been studied in depth. This paper aims to propose a customer-centric model for the study of the balance between customer-driven and technology –driven mPayment adoption. The model builds on and complements existing models and frameworks found in prior research and industry reports [1-19] and can be used to investigate directions for increasing mPayment adoption levels.

The paper is organised as follows: The next section provides information about the structure of the mPayment market and identifes the main stakeholder groups. The section following briefly reviews some mPayment models and proposes a customer-centric model. The implications for future research and development are discussed in the concluding section of the paper.

2   The mPayment Market

An mPayment transaction involves a customer, an entity offering a mobile service (called further Mobile Business Service Provider – MBSP), and the MNO who facilitates the transaction across all stages and may be actively engaged both in authorisation and completion. Banks and financial institutions (BFIs) may also be involved in transaction authorisation and completion. The MPSP participates in mPayment scenarios based on a cooperative business model. Auxiliary participants (e.g. a Mobile Internet Services Provider) may also be involved.

The players listed above interact both with each other and with the customer to facilitate an mPayment transaction from initiation to completion. As a result, the mechanisms for sharing the revenue stream become complex and costly, requiring the development and adoption of standards and protocols across the mPayment market. This may in turn lead to the diminishment of the mPayment value proposition and inhibit growth [4], [8].

2.1 Mobile Payment Market Growth and Segmentation

At present the mPayment market is dominated by Japan (close to 80%), however a global growth up to US $150 billion in 2012 with transaction revenues up to US$ 37.1 billion [9], [12]. According to [9], three particular mPayment market segments will exhibit growth in the future: mobile contactless payments, online shopping, and money transfers.

Mobile contactless payments are expected to grow from US$3 billion in 2007 to US$52 billion in 2012. These micro- or macro-transactions are conducted through a mobile payment system, developed by an MPSP and based on an alliance, or another cooperative revenue sharing formation among industry players. Initiatives such as "Pay-Buy" and "Payez Mobile" for example require collaboration among MNOs, BFIs, mobile phone manufacturers, MBSPs, and customers) [6-7]. Geographically, contactless payments at the point of sale are expected to grow across the European Union (EU), the USA, Japan, and in some Asia-Pacific countries [5 - 7], [12].

Mobile money transfers are expected to grow from US$1 billion in 2007 to US $58 billion in 2012). These transactions involve BFIs and possibly additional stakeholders such as local convenience store owners who may act as cash providers to customers in a remote area in a developing country. Mobile transfers are normally remote macro-payments, settled via an account at a bank and/or at an intermediary.

Mobile online shopping (which requires access to the Internet via a mobile device) is a basic eCommerce business-to-consumer model in which the customer accesses the Internet via a handheld device connected to a MNO. The expected growth is from US$8 billion in 2007 to US$41 billion in 2012. Both micro- and macro –payments can be made. Mobile ISPs are involved as intermediaries.

The highlighted trends emphasise growth across the spectra of transaction type, technology and settlement method, indicating that all mPayment market players involved in creating and offering mPayments services will continue to be active. The next subsection groups the mPayment market players into main stakeholder groups.

2.2 Mobile Payment Stakeholder Groups

mPayment services and systems require a high level of interoperability and compatibility across devices, network platforms, and software applications; depending on their market role, stakeholders may operate independently or participate in cooperative models. The mCommerce framework proposed in [16] can be used to identify the main stakeholder groups with regard to their role in driving the processes of spread and adoption of mPayment. Three different categorise emerge (Table 1): Primary mPayment providers (mPayment Technology Enablers, or MPTEs), secondary mPayment providers (mPayment Service Enablers, or MPSEs), and mPayment adopters (MPAs).

Table 1. The main stakeholder categories involved in mPayment development, deployment and adoption

Primary mPayment Providers: Technology Enablers (MPTEs)	Secondary mPayment Providers: Service Enablers (MPSEs)	mPayment Adopters (MPAs)
• Mobile Network Operators (MNOs) • Mobile Subscription Service Providers (MSSPs) • Banks /financial institutions (BFIs) • Mobile technology developers • Mobile device manufacturers	• Mobile content developers • Mobile content aggregators • Business and organizations accepting mobile payment (MBSPs) • Intermediaries (mobile payment aggregators, security providers) • Mobile Payment Solution Providers (MPSPs)	• Mobile commerce customers and users of new mobile content services • Customers of traditional services using mPayment methods • Customers using mobile versions of existing services (bundled with an mPayment method)

The first group includes MNOs, MSSPs, BFIs, mobile technology developers, and mobile device manufacturers. These stakeholders enable technologically mPayment and are posed to benefit directly from each mPayment transaction. Therefore, MPTEs may be considered as playing a fundamental role in the development of mPayment services, and driving the mPayment market based on technological development, often through industrial alliances [1], [3-8], [10], [12-13].

The second stakeholder group comprises market players who provide mPayment as either a service bundled with an mCommerce service, or as a payment mechanism for other goods and services [11]. More specifically, participants in the MBSPs subgroup include: merchants and organizations using mPayment at the point of sale (POS) or remotely, either for an existing product or service such as paying for a bus ticket using a mobile phone, or for a new 'pure mobile' service (such as a mobile game download). MBSPs benefit from retaining their existing customer base and from generating new revenue streams through innovative services. Intermediaries (for example, security and identification providers, and payment aggregators) and MPSPs are businesses which provide mPayment services through their core business model [4-5], [8], [18]. The MPSE stakeholders drive the mPayment market based on business development: mPayment services are deployed in their respective business models because of the technological drive of the primary mPayment providers.

The third category includes customers and end-users who participate by adopting mPayment. Studies about customer motivation have identified motivational and decision making factors, and critical success factors and barriers to acceptance and adoption. The customer subgroups in the table refer to the target groups of mCommerce customers, and to customers who use contactless payment for traditional or innovative services [11]. Customers have been found to be willing to accept mPayment services depending on the context of the offer but have not created a strong demand for them [8-9], [11], [13-15], [17].

It may be concluded that mPayment at present is still technology- rather than cus-tomer-driven, mostly through initiatives by the primary mPayment providers. Service-driven mPayment occurs in areas such as new mobile content services and applications (e.g. mobile entertainment [21]), and 'mobile versions' of existing services (for exam-ple, mobile betting [22]). Despite opinions such as "In our contemporary society, being mobile, or simply capable of playing with mobility options, thanks to adequate infra-structure, devices, skills and knowledge, is generally associated with a positive, dy-namic and seemingly indispensable form of lifestyle as well as productive behaviour" [23, p. 79], customers are yet to create a strong demand for mPayment services.

3 Modelling mPayment

A number of models have been used in the literature to represent the supply side of mPayment and identify the value proposition of mPayment. Value chain models, ser-vice models (scenarios), and business models involve the players in the technology enablers and service enablers stakeholder groups [1-7], [10], [12-13], [25]. The de-mand side (customers and end-users) has been modelled through acceptance and adoption models to identify critical success factors and barriers to adoption [11], [14-17], [25-30]. This section reviews the relevant findings and proposes a model for the study of the dynamic relationships between the stakeholder groups and between mPayment adopters and stakeholders.

3.1 Supply Side Studies: mPayment Value Chain and Scenarios

The value chain approach has been used to identify the players involved in certain mPayment scenarios, and their roles [1], [19]. To capture the complexity of the inter-actions, using a web of value chains rather than a linear representation has been also suggested [24].

The interactions between the players have been modelled using scenarios and use cases [13], [25]. A set of seven disjoint use cases identified in prior work was used in [13] to derive the important characteristics of the mPayment value proposition with respect to customers: geographical applicability, payment guarantee, mobile market integration, and payment amount.

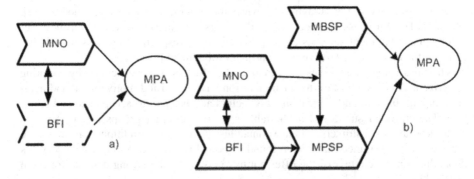

Fig. 1. The evolution of mPayment (based on [9-10])

A dichotomy of 'carrier-centric' and 'payment solution-centric' models of mPayment scenarios is proposed in [9]. Their analysis of the development of mPayment services across the market place indicates that mPayment has evolved from a stage where MNOs act also a MBSP and MPSP (Figure 1a) to a stage where mobile business services are unbundled from mobile data, and MBSPs have started to drive the mPayment market (Figure 1b).

3.2 mPayment Adoption and Acceptance

As summarised in [19] the demand side of mPayment (customers and end-users) has been studied empirically and qualitatively through models such as TAM (Technology Acceptance Model) and TTF (Task-Technology Fit), including participants from a country or a region [3], [8], [11], [14-15], [25-30]. These studies have identified a number of mPayment factors influencing negatively or positively customer adoption and use (Figure 2).

Cost, convenience, and added value were repeatedly identified as critical mPayment success factors. **Direct cost** (customer paying an additional charge to use mPayment in a specific scenario and customer paying for mobile data transport) has been shown to be a barrier to the intention to use, also because of the abundance of other, less costly methods already available. The **enabling cost** (customer needing a mobile device supporting the technology used for mPayment) may also become a barrier but its importance may vary depending on the customer demographics [2], [8], [14], [26-27]. **Convenience** (related to 'perceived ease of use' in TAM) refers to the degree of effort needed by the customer to execute a payment (including registration, access, device usability, time needed to complete the transaction) [2], [8], [11], [14-15], [17]. **Value added** (related to 'perceived usefulness' in TAM) refers to the additional benefits for the customer when using mPayment such as saving time, saving the need to interact at POS, replacing the need to carry cash or use multiple plastic cards) [8], [11], [14-15], [17], [27].

Four additional characteristics have been identified: Mobility support, task-technology fit, trust and security. Customer **mobility support** refers to the ability of the customer to use mPayment across geographical locations, including internationally, and in use situations where mPayment becomes the only viable method (i.e. the ability to transact not only 'any time' but 'anywhere')[14], [17]. **Task-technology fit** refers to the extent to which mPayment technical features match both the customer ability to operate the device in order to pay, and the suitability of mPayment for the particular service or product [11]. **Security** and **trust** refer to the perceptions of the customer with regards to the non-repudiation of the mPayment transaction [11], [14]. The importance if these factors have not been established with a great degree of certainty. While mobility support and task-technology fit have been found to influence positively the intention to use, security and trust have not been found to be critically important.

Three areas of customer demand emerge from the discussion of he critical success factors and barriers to mPayment: a) Demand for quality of service (convenience, value added, mobility support, task-technology fit; b) Demand for cost-effectiveness (direct and indirect costs), and c) Demand for a regulated environment (security and trust). Quality of service plays a critical role in customer acceptance, however the

Fig. 2. mPayment characteristics influencing customers and end-users

decision to adopt is a tradeoff between value and cost. The environment in which mPayment occurs is normally perceived to be trustworthy, partially because customers have already established relationships with some of the players (e.g. with the MNO).

3.3 A Customer-Centric Demand and Supply Model

As mentioned earlier, mPayment adoption has not progressed according to the forecasts in the past. In the previous subsection MNOs and possibly BFIs were identified as drivers at the initial stage of mPayment; MPSPs become the main driver of the market as mPayment develops further, with MBSPs also becoming active as they start their own adoption process. In [11] the authors note, "Mobile payments represent an extremely interesting paradox in the world of mobile telecommunications, still not showing success in most markets. Customer acceptance turned out to be a decisive factor". Therefore placing an emphasis on customers as potential drivers of mPayment may bring a new perspective to the study of mPayment development and spread.

As customer acceptance is critical to the success of mPayment, the two other stakeholder groups need to develop and offer services meeting the demands of the MPA stakeholder group identified above: An mPayment service of high quality and at a cost which the customer will be willing to pay. The MPTE/MPSE stakeholders need to engage in cooperative business models for service provision and revenue sharing which will allow meeting customer demand and remaining viable. Figure 3 shows a demand and supply model which can be used to study the dynamics of these processes.

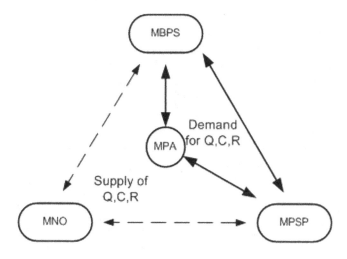

Q- Quality, C- Cost-Effectiveness, R- Regulated Environment

Fig. 3. A Customer-centric demand and supply model

MPAs demands for quality, cost- effectiveness and a regulated environment are directed mainly towards the MBPS participants. However, it is the relationships among all industry players (MNOs, MPSP and MBPS) which underpin both the value and the quality parameters. The demand for a regulated environment is met primarily by the MNO and the MPSP, operating within a local or a regional regulatory framework. The model identifies three main areas of further investigation:

1. How will MBSPs meet customer demand in terms of quality and cost-effectiveness? How can customer demand be more accurately predicted and what new services requiring mPayment or bundled with an mPayment option could be developed to satisfy them?
2. How will MPSPs meet customer demand in terms of quality and cost-effectiveness in a regulated environment? How could mPayment become the preferred mode of payment for customers (presuming this is desirable from a customer perspective)?
3. How will MNOs (and mobile technology providers) support the development of quality mPayment services in cooperation with MBSPs and MPSPs?

In the next section, an mPayment case study is analysed by applying the model to answer the questions above.

4 The 'TXT-a-Park' Case Study

Customers in some city council owned car parks in New Zealand cities have the choice of paying for their parking by coin, by credit card or by SMS (text messaging). The latter service is known as 'TXT-a-park'. After the SMS-based transaction is completed, the customer receives a ticket from the parking meter. The cost of parking is debited from the mobile network subscriber account [31-33].

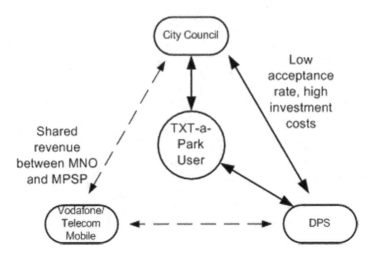

Fig. 4. The 'TXT-a- park' case study customer-centric model

Figure 4 shows the supply and demand model for TXT-a-park. MNOs are the two dominating New Zealand mobile operators Vodafone and Telecom Mobile. The MBSP is the relevant City Council. The MPSP is DPS – an established New Zealand eCommerce payment solution developer and provider company.

The double arrowed lines in the model show the relationships amongst the model participants. The city council collaborates with the mobile network operators, on one side, and with the payment solution company on the other, in order to set up and offer the service. The payment solution provider facilitates mobile (and credit card) payment and provides and supports the revenue channels for of all participants. The customer pays a regular service cost (parking fee) to the city council. The direct mPayment costs include a surcharge on top of the normal text message cost; the surcharge is shared by the network operator and the payment solution provider.

According to the claims of the mobile service provider, customer demand in terms of quality has been met as the mPayment service is perceived as convenient, and useful [31]. However, the service does not support mobility as the transaction has to be conducted in the proximity of the payment terminal. There is some evidence to indicate that the cost of the transaction may be perceived as too high by some customers as the level of use seems to vary according to the location of the parking area. This was found in the course of a small project conducted by the author and a student who observed several parking meters and briefly interviewed customers at the spot [34]). It may be surmised that in areas that are more affluent customers are less sensitive towards the extra charge. The transaction environment is perceived as trustworthy: customers trust inherently the MBSP and are aware of the security provisions of the two network operators. In summary, the quality of this service (Q) is high and the regulated environment (R) is supportive of the service however the cost-effectiveness (C) may be low for some customers.

The analysis of the case allows providing some answers to the questions formulated in Section 3.

1. With respect to MBSPs meeting customer demand in terms of quality and cost-effectiveness, new valued added services could be offered – for example, extending remotely the validity of the parking ticket remotely as suggested in [12].
2. Increasing the scope of mPayment options by adding payment via a bank account (already implemented on a trial basis for business accounts [34]) addresses the second question - about MPSPs meeting customer demand and increasing the appeal of mPayment as a mode of payment. It may be expected that customers parking while on business duty will be highly motivated to pay direct and have to go through a laborious claim process afterwards.
3. Referring to the question about MNOs (and mobile technology providers) supporting the development of quality mPayment services in cooperation with MBSPs and MPSPs: In this case, the MNOs and the MPSP have negotiated, in return for collaborating with the mobile service provider, to share the total surcharge revenue. As the MNOs are satisfied with their return on investment, it is unlikely that they would come up with a new initiative to promote mPayment (as the mPayment service is not an essential part of their business model).

The brief discussion of the TXT-a park case study demonstrated the applicability of the model proposed earlier to the analysis of an mPayment service: The relationships were substantiated and the questions formulated were addressed. However as only limited data were used conclusions about the generalizability of the customer-centric model cannot be made without a further study.

5 Concluding Remarks

Prior research has addressed some of the relationships in the supply and demand areas in the customer-centric model proposed. The ability of the MNO to provide service matching customer demand has been studied in [17] where the issue of balancing customer needs and the business value proposition of one of the main players in the MPTE stakeholder group is investigated. The study used the critical success factors identified in earlier work to analyze how well wireless technologies would meet customer user requirements. Balancing the needs of MBSPs and the drive by MNOs and MPSPs is the focus of [18], where four barriers preventing MBSPs from proactively driving the mPayment market are identified: relative advantage, compatibility, complexity and cost. The customer-centric model introduced and validated here aligns well with the direction of these studies and also with a number of the suggested general mPayment research directions in [19]. It is proposed to use the model to investigate and evaluate mPayment services within the framework of customer requirements, in order to suggest to MBSPs directions for further development meeting customer mobility and lifestyle needs and involving customers as active participants. Further studies may adopt both a quantitative approach to study the perceptions, attitudes and needs of customers, and a qualitative approach to study the needs of MBSPs. A set of measures to evaluate and assess the parameters of the model (quality of service, cost-effectiveness and regulatory environment) will need to be developed.

Acknowledgements

The author would like to thank the anonymous reviewers for the thorough reviews and helpful suggestions and comments.

References

1. Mobile Payment Forum 1,
   `http://www.mobilepaymentforum.org/pdfs/mpf_whitepaper.pdf`
2. Kreyer, N., Pousttchi, K., Turowski, K.: Standardized Payment Procedures as Key Enabling Factor for Mobile Commerce. In: Quirchmayr, G., Tjoaa, A.M. (eds.) 3rd International Conference on Electronic Commerce and Web Technologies, pp. 400–409. ACM Press, New York (2002)
3. Valcourt, E., Robert, J.-M., Beaulieu, F.: Investigating Mobile Payment: Supporting Technologies, Methods, and Use. In: IEEE International Conference on Wireless and Mobile Computing, Networking and Communications, pp. 29–36. IEEE Press, New York (2002)
4. Ondrus, J., Pigneur, Y.: A Disruption Analysis in the Mobile Payment Market. In: 38th Annual Hawaii International Conference on System Sciences, pp. 10–84. IEEE Press, New York (2005)
5. Mobile Payment Forum 2, `http://www.mobilepaymentforum.org/documents/Proximity_Payment_IR_11_0.pdf`
6. Payez Mobile (2007), `http://www.gemalto.com/press/IntheNews/download/2007/11-09-2007-payez_mobile.pdf`
7. GSMA, `http://www.gsmworld.com/news/press_2007/press07_21.shtml`
8. Dahlberg, T., Mallat, N.: Mobile Payment Service Development – Managerial Implications of Consumer Value Perceptions. In: 9th European Conference on Information Systems, pp. 649–657 (2002)
9. Van Bossuyt, M., Van Hove, L.: Mobile Payment Models and Their Implications for NextGen MSPs. J. of Policy, Regulation and Strategy 9(5), 31–43 (2007)
10. Tong, F., Zhou, X., Liu, S.: The Value Chain of Mobile E-payment. In: International Conference on Electronic Commerce, pp. 880–882. ACM Press, New York (2005)
11. Pousttchi, K., Wiedemann, D.G.: What Influences Consumers' Intention to Use Mobile Payments? In: LA Global Mobility Round table (2007)
12. Taga, K., Karlsson, J., Arthur, D.: Little Global M-Payment Update (2005), `http://www.3mfuture.com/articles_epayment/Global_M-Payment-Report_Update_Arthur_D_Little_2005.pdf`
13. Pousttchi, K., Schiessler, M., Wiedemann, D.G.: Analyzing the Elements of the Business Model for Mobile Payment Service Provision. In: 6th International Conference on Mobile Business, p. 44 (p. 8). IEEE Press, New York (2007)
14. Pousttchi, K.: Conditions for Acceptance and Usage of Mobile Payment Procedures. In: 2nd International Conference on Mobile Business, pp. 201-210. University of Munchen (2003)
15. Teo, E.: Inhibitors and Facilitators for Mobile Payment Adoption in Australia: A preliminary study. In: 4th International Conference on Mobile Business, pp. 663–666. IEEE Press, New York (2005)
16. Petrova, K.: Mobile Commerce Applications and Adoption. In: Khosrow-Pour, M. (ed.) Encyclopedia of e-Commerce, e-Government and Mobile Commerce, pp. 771 – 776. Idea Group Reference, Hershey, PA (2006)

17. Zmijewska, A.: Evaluating Wireless Technologies in Mobile Payments — A Customer Centric Approach. In: 4th International Conference on Mobile Business, pp. 354–362. IEEE Press, New York (2005)

18. Mallat, N., Tuunainen, V.K.: Merchant Adoption of Mobile Payment Systems. In: 4th International Conference on Mobile Business, pp. 347–353. IEEE Press, New York (2005)

19. Dahlberg, T., Mallat, N., Ondrus, J., Zmijewska, A.: Mobile Payment Market and Research – Past, Present and Future. In: Helsinki Global Mobility Roundtable (2006)

20. Poulbere, V.: Mobile payment: progressing towards large–scale deployments. Ovum report (2008)

21. Schubert, P., Hampe, J.F.: Mobile Communities: How Viable are Their Business Models? An Exemplary Investigation of the Leisure Industry. Electronic Commerce Research 6(1), 103–121 (2006)

22. Petrova, K., Parry, D.: Mobile Computing Applications in New Zealand. In: Yoo, Y., Lee, J.-N., Rowley, C. (eds.) Trends in Mobile Technology and Business in the Asia-Pacific Region, pp. 153–177. Chandos Publishing, Oxford (2008)

23. Rossel, P., Finger, M., Misuraca, G.: Mobile e-Government Options: Between Technology-driven and User centric. The Electronic Journal of e-Government 4(2), 79–86 (2006)

24. Akkermans, J.M., Baida, Z., Gordijn, J., Peña, N., Altuna, A., Laresgoiti, I.: Value Webs: Using Ontologies to Bundle Real-World. Intelligent Systems 19(4), 57–66 (2004)

25. Herzberg, A.: Payments and Banking with Mobile Personal Devices. Communications of the ACM 46(5), 53–58 (2003)

26. Dahlberg, T., Mallat, N., Oorni, A.: Consumer Acceptance of Mobile Payment Solutions – Ease of Use, Usefulness and Trust. In: 2nd International Conference on Mobile Business, pp. 211-218 (2003)

27. Malat, N.: Exploring Consumer Adoption of Mobile Payments - A Qualitative Study. In: Helsinki Global mobility Round table (2006)

28. Hort, C., Gross, S., Fleish, E.: Critical Success Factors of Mobile Payments, In: M-Lab (2002), http://www.m-lab.ch/pubs/13_CriticalSuccess_Mobile

29. Wrona, K., Schuba, M., Zavagli, G.: Mobile Payments - State of the Art and Open Problems. In: Fiege, L., Mühl, G., Wilhelm, U.G. (eds.) WELCOM 2001. LNCS, vol. 2232, pp. 88–100. Springer, Heidelberg (2001)

30. Henkel, J.: Mobile Payment: The German and European Perspective. Mobile Commerce 10(2), 1–22 (2001)

31. Auckland City Council, http://0-www.aucklandcity.govt.nz.www.elgar.govt.nz/news/releases/20051214a.asp

32. Wellington City Council, http://www.wellington.govt.nz/services/parking/onstreet/txtapark.html

33. Fronde, http://www.frondeanywhere.com/txt-a-park/

34. Mehra, R.: Paying for parking with Your Mobile Phone, unpublished

Ontological Vulnerability Assessment

Aaron Steele

School of Engineering and Advanced Technology, Massey University,
Palmerston North, New Zealand
A.Steele@massey.ac.nz

Abstract. Vulnerability assessment is a vital part of the risk management process. The accuracy and reliability of calculated risk depends on comprehensive and correct assessment of system vulnerabilities. Current vulnerability assessment techniques fail to consider systems in their entirety and consequently are unable to identify complex vulnerabilities (i.e. those vulnerabilities that are due to configuration settings and unique system environments). Complex vulnerabilities can exist for example when a unique combination of system components are present in a system and configured in such a way that they can be collectively misused to compromise a system.

Ontologies have emerged as a useful means for modeling domains of interest. This research shows that taking an ontological approach to vulnerability assessment results in improved identification of complex vulnerabilities. By ontologically modeling the domain of vulnerability assessment, the resulting ontology can be instantiated with a system of interest. The process of instantiating the ontology doubles as a technique for methodically discovering complex vulnerabilities present in the given system. Furthermore, it is suggested that the instantiated ontology will also be able to be queried in order to discover additional complex vulnerabilities present in the system by reasoning through implicit knowledge captured by the instantiated ontology.

1 Introduction

Vulnerability assessment is a vital part of the risk management process [14]. The accuracy and reliability of calculated risk depends on comprehensive and correct assessment of system vulnerabilities. Current vulnerability assessment techniques fail to consider systems in their entirety and consequently are unable to identify complex vulnerabilities (i.e. those vulnerabilities that are due to configuration settings and unique system environments) [8,16]. Complex vulnerabilities can exist for example when a unique combination of system components are present in a system and configured in such a way that they can be collectively misused to compromise a system.

Ontologies have emerged as a useful means for modeling domains of interest [7]. This research shows that taking an ontological approach to vulnerability assessment results in improved identification of complex vulnerabilities. This is achieved by ontologically modeling the domain of vulnerability assessment. The

S. Hartmann et al. (Eds.): WISE 2008, LNCS 5176, pp. 24–35, 2008.

resulting ontology can then be instantiated with a given system of interest. The process of instantiating the ontology doubles as a technique for methodically discovering complex vulnerabilities present in the given system. Furthermore, it is suggested that the instantiated ontology will also be able to be queried in order to discover additional complex vulnerabilities present in the given system by reasoning through implicit knowledge captured by the instantiated ontology.

2   Background

2.1   Risk Management

The risk management process aims to identify and control system security risks. Typically, the overall process has the following structure [13,14].

1. Asset Identification
2. Vulnerability Assessment
3. Threat Assessment
4. Risk Assessment
5. Implementation of Countermeasures

Firstly valuable system assets are identified. Next vulnerabilities and threats to those valuable assets are identified. Then the actual risk is assessed (e.g. vulnerabilities prioritised by most likely or most damaging etc). Finally, what are deemed as appropriate countermeasures are then implemented to control the vulnerabilities in order of priority in the hope of securing the system by minimizing or eliminating the risk.

Accurate and comprehensive vulnerability is vital for the overall process to work. If some vulnerabilities are not identified then no risk is associated with them, and therefore no countermeasures are implemented, resulting in an otherwise secure system potentially being left wide open for attack.

2.2   Current Vulnerability Assessment Techniques

Current vulnerability assessment techniques fall into two main categories: Evaluation against what is already known, and utilisation of human resources.

Evaluating a system against already known vulnerabilities is a very common and useful vulnerability assessment technique [17]. At its core this technique has the following philosophy: Compare all of the components of a system (e.g. the different software packages installed on a system) with a vulnerability database or repository to see if any of the existing system components have known vulnerabilities [4,17]. This comparison can be done manually (e.g. finding out the components installed on a system and then searching for them on an online vulnerability database such as nvd.nist.org) or it can be automated by using an vulnerability scanner (e.g. Nessus which has a built in repository [18]). This technique can be very useful and save a lot of time and effort. On the downside this technique, has been known to provide false positives (e.g. a scanner report a vulnerability that doesn't actually exist) and false negatives (e.g. a scanner missing vulnerabilities that do exist) [3,12]. Furthermore, the vulnerabilities that this technique identifies are always package specific. Vulnerabilities that result from a combination of system components being used together to violate a system specific security goal, for example, cannot be identified using this technique.

The other category of vulnerability assessment aims to utilise the human resources of a given system. At its core this technique has the following philosophy: Ask the users, developers, and administrators of the system in question to identify what they think are vulnerabilities in the system [13,14]. The idea is to use the people who are the most familiar or experienced with the system to discover vulnerabilities that they may have come across in their course of use. This technique can be aided by the use of a matrix that maps system assets against security attributes [1]. Although this technique can occasionally identify more complex vulnerabilities, it falls short of being a comprehensive system assessment. Furthermore, it relies almost entirely on the experience of the system users.

Even with both categories of techniques being used together to assess a system complex vulnerabilities are often missed. These complex vulnerabilities can potentially leave a system wide open to attack. The question that emerges is: how can a system be methodically assessed for complex vulnerabilities?

2.3 Ontologies

Recently ontologies have emerged as a useful way of modelling a domain of interest [7]. As a relatively new discipline the amount of research focused on ontologies in information systems security is limited, let alone vulnerability assessment. Nevertheless the literature provides some interesting insights into the usefulness of ontologies. Research shows that there is potential for concepts such as vulnerabilities to be modelled at different levels of abstraction using ontologies [10]. It also shows that vulnerabilities can be categorised in such a way that a division can be seen between those due to technology and those due to higher level concepts such as social engineering or policy oversight [11]. Furthermore, it shows that concepts involved in the security domain can be organized by hierarchy and that the relationships between those concepts can also be modelled [15]. The literature gives example of how an ontology could be used as a methodical tool [2,5]. It also gives an example of the phases involved when instantiating a security ontology from an arbitrary system [19,20]. Finally examples can be seen of how ontologies can be queried to answer certain questions [6,9].

Given the above question: how can a system be methodically assessed for complex vulnerabilities? The proposed answer becomes: by using an ontology.

3   Vulnerability Assessment Ontology

The first step in proving an ontology would be useful to solve this problem was to first ontologically model the domain of vulnerability assessment. In order to achieve this a number of resources needed to be considered. These resources included; other existing ontologies that involved concepts similar to those involved in vulnerability assessment, the vulnerability assessment domain itself, vulnerability assessment literature, and ontology development literature. The main question during this process was what are the underlying characteristics of vulnerable systems? That is, what makes a system vulnerable on a principle level? The answer stemmed from a humorous quote by Dennis Hughes of the FBI which reads, "the only truly secure computer is one that's unplugged, locked in a safe, and buried 20 feet under the ground in a secret location and I'm not even too sure about that one". Although secure because of its inaccessibility, a computer in this state is obviously unusable. The key to the answer began to emerge as access points. For a system to be secure every single way it is accessed needs to be controlled or secured. The more access points a system has the higher the chance that one of those access points may be uncontrolled or unsecured. Therefore to check if a system is vulnerable, one needs to first discover the access points of the system and then see if they are controlled. The next step was then an iterative process where the developed ontology was instantiated against real world systems. The resulting instantiated ontology was then used to revise the underlying ontology (i.e. removing redundant classes or relationships, adding new classes or relationships, and refining existing classes and relationships). A graphical representation of the vulnerability assessment ontology in its current form can be found in figure 1.

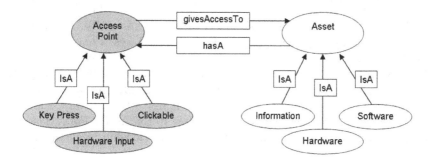

Fig. 1. Graphical representation of the Vulnerability Assessment Ontology

The vulnerability assessment ontology has two main parts. The first is the Access Point class, an access point can be thought of as any part o the system that allows user input which in turn causes the state of the system to change. The Access Point class has three main subclasses as indicated by their IsA relationship. These subclasses in essence represent the three main inputs of a generic computer. The first is the Key Press class. This class encapsulates all and every keyboard key combination. These include function keys (e.g. F1, F2, F3, etc), hotkey combinations (Ctrl + S, Ctrl + Shift + D, etc), text box inputs (e.g. username, password), or any other key press that causes the system to change state.

The second is the Clickable class. This class represents anything that can be clicked by the pointing device (i.e. mouse, touchpad, etc). These include single, double and triple clicks, of every available button (e.g. left mouse button, right mouse button, scroll wheel, etc). The third is the Hardware Input class. This class represent parts of the system that allow hardware inputs. This practically speaking exists to represent mainly USB ports and CD/DVD drives, but also any other hardware input. Although a CD drive for example can be considered a system asset, it is here also classified as an access point. This is because there are situations where a CD drive may exist but have no access point (e.g. it is locked in a cabinet with the rest of the computer). Therefore the access points of these hardware assets have been included as a unique subclass of access point. The second part of the ontology is the Asset class. An asset is usually defined as any part of a system that has some value worth protecting. For the purpose of this ontology this definition will be used but also extended to not only include parts of a system with some value, but also parts of a system that can be used to reach another part of a system with value. The three subclasses of Asset are: Software, Information, and Hardware. Software includes not only specific software applications, but also subparts of applications (e.g. 'Notepad.exe' is an asset, but the 'Open file' dialogue box in Notepad is also considered an asset). Information typically represents data files of various types (e.g. text files, image files, database files, proprietary file types, etc). Finally, hardware assets include drives (e.g. disk, USB, CD/DVD, etc), peripherals (e.g. scanners, printers, cameras, etc), and any other pieces of potentially useful hardware (e.g. modems). The last part of the ontology to explain is the two relationships that exist between the Access Point class and the Asset class. The first relationship goes from the Access Point class to the Asset class and is called givesAccessTo. As the name indicates this relationship represent how a particular Access Point gives access to a particular Asset. A simple example would be pressing the Windows button on the keyboard gives access to the start menu. In this example the Windows button is an Access Point of subclass Key Press with a givesAccessTo relationship to the start menu which is an Asset belonging to the Software subclass. The second relationship goes from the Asset class to the Access Point class and is called hasA. Again, as the name suggests, this relationship represents how a particular Assets has an Access Point. Continuing with the above example the start menu has a number of clickable icons, one of which is the Run icon. In this situation the start menu is again an Asset belonging to the Software subclass. It however also now has a relationship of type hasA with the Run icon which is an Access Point of the Clickable subclass.

4   Proof of Concept

As a proof of concept the vulnerability assessment ontology was instantiated against a New Zealand university's library catalogue computers. The objective of these computers is to only allow user access to the library catalogue. The catalogue is presented via a web browser and allows users to anonymously search the catalogue and also allows student and staff users the ability to log in to view their current lending information. These are the only functions that these

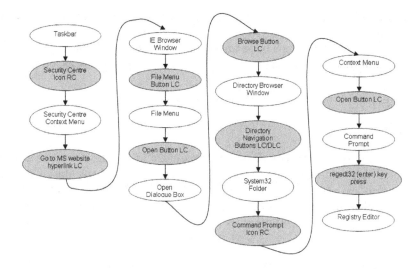

Fig. 2. Taskbar Access Chain via Security Centre System Tray Icon

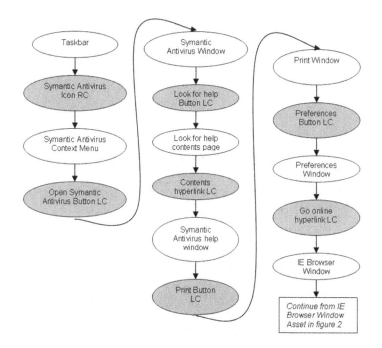

Fig. 3. Taskbar Access Chain via Symantic Antivirus System Tray Icon

computers should provide to the user. These security requirements state what functionality these computers should allow. The aim of instantiating the ontology is to discover in a methodical fashion what functionality these computers actually do allow.

The first step in instantiating the ontology is to identify the initial assets directly accessible to the user when the system is in its initial state (e.g. usually what's on the screen after the system has been turned on and has finished loading, and any hardware inputs). The initial assets of the catalogue computers are as follows:

– Taskbar
– CD Drive
– USB Port
– Browser Window

Of the four initial assets, two are software assets, and two are hardware assets. The second step is to identify the access points of each of the initial assets. Recalling that an access point is defined as a part of the system that allows user input which in turn causes the state of the system to change, it is therefore necessary to test each potential access point (e.g. try different click types etc).

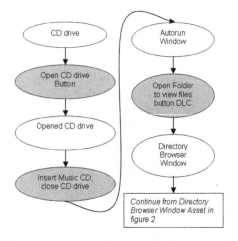

Fig. 4. CD Drive Access Chain

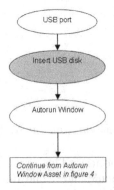

Fig. 5. USB Drive Access Chain

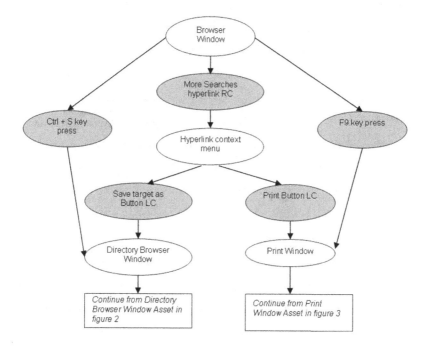

Fig. 6. Browser Window Access Chain

To begin the taskbar will be assessed (Note: LC = Left Click, RC = Right Click, DLC = Double Left Click, MO = Mouse Over). The access points of interest provided by the taskbar are as follows:

- Taskbar
  - Security Centre system tray icon RC
  - Symantic Antivirus system tray icon RC

Now given that each access point changes the state of the system, the new state of the system needs to be reassessed to discover any new assets, and in turn new access points, which in turn will cause the process to be repeated. Eventually a chain linking assets to access points is developed. This chain gives a more accurate picture of the true functionality offered to the user. Figures 2 - 3 show the chain after further enumeration of assets and access points. Note: the three diagrams begin with the same asset Taskbar in its initial state but have been split into two diagrams due to space restrictions. For clarity the Asset bubbles are white and the Access Point bubbles are grey. Some detail of the instantiated ontologies have been omitted due to space restriction that occur when representing ontologies graphically. These omissions include the explicit declaration of Asset or Access Point subclasses, and the labelling of the relationships between Assets and Access Points. However all relationships flowing from an Asset to an Access Point will always be of type hasA and all relationships flowing from an Access Point to an Asset will always be of type givesAccessTo.

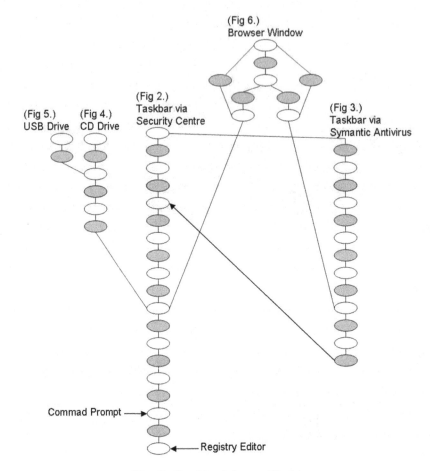

Fig. 7. Combined Access Chains

The next step is to repeat the same process with the remaining initial assets. Figures 4 and 5 represent the access chains of the CD drive, and USB port assets. Figure 6 represent the access chain for the Browser window. Finally Figure 7 shows the access chains from Figures 2-6 combined in order to show the overall interlinking of the access chains (note: the flow of access in Figure 7 is always downwards except when otherwise indicated by an arrow). Due to space restrictions only the starting points of each access chain are indicated in Figure 7, along with two final valuable assets. For more detail of Figure 7 please refer to Figures 2-6.

5 Summary

From the four initial assets eight different chains have been found that actually provide the user a way to access the file system including valuable assets such as the command prompt and the registry editor (not to mention the file system as a valuable asset itself). Upon closer examination the default user profile also has write access to many parts of the file system, including the start up folder. This is of particular concern given that malicious software (e.g. trojans, viruses, or key loggers) could be copied from external devices such as a USB drive or CD which have also been shown to available to the user. Also with a valid student login the internet explorer window can be used to access the entire internet. As a side note, in what is a bizarre twist of intention, two of the system tray icons which actually exist as security controls (i.e. security centre and symantic antivirus) can actually be used on the path to compromising the system. These are all prime examples of complex vulnerabilities, none of which are due to programming errors, or unpatched software, but are all due to configuration settings and the unique system environment (i.e. internet explorer as the default web browser, the print preferences offered by a unique printer which includes a hyperlink, autorun menu for USB and CD allowed by default, etc).

An interesting observation from instantiating the ontology is that of information reuse. Even in this small example it can be seen when an asset is reached that has already been instantiated, re-instantiation is not required, but the previous instantiation can be reused, including all subsequent access points and assets. This again has an interesting consequence. In viewing the final instantiated ontology key assets and access points can be identified (i.e. assets and access points that get used by multiple chains) for example the print menu asset in the above example. These key assets and access points then become a logical place to fix the vulnerabilities. For example if asset X is used by seven different chains to compromise the system, instead of addressing every chain individually, removing asset X may be the most cost effective option. Although this stream of thought is swaying more away from vulnerability assessment and more toward risk and cost calculations. Nevertheless, this proof of concept has shown that the use of this ontology can help discover in methodical fashion complex vulnerabilities that exist in a given system.

6 Future Work

The ontology in its current form is showing some exciting results, however this research is not yet entirely complete. Future goals of this research include the following:

- Incorporation of security attributes (i.e. the CIA triad or extension) into the ontology.
- Instantiation of the ontology with different types of systems (i.e. web based systems).
- To validate the usefulness of the ontology, the vulnerabilities found using the ontology will be compared against the vulnerabilities found using traditional assessment techniques.
- Using Protégé (an ontology development environment) the ontology will be formally written in OWL and instantiated. This will also allow for automated reasoning and querying (functionality also offered by Protégé).

– Analysis of reverse scans of access chains (i.e. from valuable systems assets back to initial assets).
– Assessment of appropriate queries and languages will also be explored in the hope of gleaning additional implicit information about complex vulnerabilities from instantiated ontologies.

References

1. Antón, P.S., et al.: Finding & Fixing Vulnerabilities in Information Systems: The vulnerability assessment & mitigation methodology. RAND National Defence Research Institute (2003)
2. Bagchi, A., Atluri, V. (eds.): ICISS 2006. LNCS, vol. 4332. Springer, Heidelberg (2006)
3. Beaver, K.: Security scan results: Take them with a grain of salt, Windows Security Tips (2006), http://searchwindowssecurity.techtarget.com/tip/0,289483,sid45_gci1227130,00.html
4. Cobb, M.: Should every flaw in a vulnerability scanner report be addressed? Ask The Security Expert: Questions & Answers. (2006), http://searchsecurity.techtarget.com/expert/KnowledgebaseAnswer/0,289625,sid14_gci1244322,00.html
5. Ekelhart, A., et al.: Security Ontologies: Improving Quantitative Risk Analysis. In: Proceedings of the 40th Annual Hawaii International Conference on System Sciences (HICSS 2007). IEEE Computer Society, Los Alamitos (2007)
6. Funabashi, M., Grzech, A. (eds.): Employing Ontologies for the Development of Security Critical Applications IFIP 2005, I3E 2005, vol. 189. Springer, Heidelberg (2005)
7. Gruber, T.R.: Toward principles for the design of ontologies used for knowledge sharing. In: Guarino, N., Poli, R. (eds.) Formal Ontology in Conceptual Analysis and Knowledge Representation, pp. 907–928. Academic Press, Inc., London (1995)
8. JNSM. Call for Papers: Journal of Network and System Management. Special Issue on Security Configuration Management (2008), http://www.mnlab.cs.depaul.edu/events/JNSM-secmgmt/
9. Karyda, M., et al.: An ontology for secure e-government applications. In: First International Conference on Availability, Reliability and Security (ARES 2006). IEEE Computer Society, Los Alamitos (2006)
10. Kim, A., Luo, J., Kang, M.: Security Ontology for Annotating Resources. In: 4th International Conference on Ontologies, Databases, and Applications of Semantics (ODBASE 2005), Agia Napa, Cyprus. Springer, Heidlberg (2005)
11. Manandhar, S., Austin, J., Desai, U., Oyanagi, Y., Talukder, A.K. (eds.): AACC 2004. LNCS, vol. 3285. Springer, Heidelberg (2004)
12. Nilsson, J.: Vulnerability Scanners, Master of Science Thesis at Department of Computer and Systems Sciences, Royal Institute of Technology, Kista, Sweden (2006)
13. Peltier, T.R.: Information Security Risk Analysis, Auerbach (2001)
14. Pfleeger, C.P., Pfleeger, S.L.: Security in Computing, 4th edn. Prentice Hall, Westford (2006)
15. Raskin, V., et al.: Ontology in information security: a useful theoretical foundation and methodological tool. In: Proceedings of the 2001 workshop on New security paradigms. ACM Press, Cloudcroft (2001)

16. Shah, S.: Detecting Web Application Security Vulnerabilities. O'Reilly SysAd-
    min   (2006),http://www.oreillynet.com/pub/a/sysadmin/2006/11/02/webapp_
    security_scans.html
17. Stoneburner, G., Goguen, A., Feringa, A.: SP 800-30 Risk Management Guide for
    Information Technology Systems, National Institute of Standards and Technology
    (2002)
18. Tenable. Nessus: The network vulnerability scanner. Accessed (February 2008),
    http://www.tenablesecurity.com/nessus/
19. Tsoumas, B., Gritzalis, D.: Towards an Ontology-based Security Management. In:
    Proceedings of the 20th International Conference on Advanced Information Net-
    working and Applications (AINA 2006). IEEE Computer Society, Los Alamitos
    (2006)
20. Tsoumas, B., et al.: Security and Privacy in Dynamic Environments. In: Fischer-
    Hubner, S., Rannenberg, K., Yngstrom, L., Lindskog, L. (eds.) IFIP International
    Federation for Information Processing, pp. 99–110. Springer, Boston (2006)

Process Model Elicitation and a Reading Technique for Web Usability Inspections

Tayana Conte[1,2], Verônica T. Vaz[1], Jobson Massolar[1], Emilia Mendes[3], and Guilherme Horta Travassos[1]

[1] PESC – COPPE/UFRJ, Private Bag 68.511, CEP 21945-970, Rio de Janeiro, RJ, Brazil
[2] Computer Science Department, Federal University of Amazonas, Av. Rodrigo Octávio Ramos 3000, CEP 69077-000 Manaus, Amazonas, Brazil
[3] Computer Science Department, The University of Auckland, Private Bag 92019, Auckland, New Zealand
{tayana,jobson,ght}@cos.ufrj.br,
veronica.taquette@poli.ufrj.br, emilia@cs.auckland.ac.nz

Abstract. Given the growth in the number and size of Web Applications worldwide, Web quality assurance, and more specifically Web usability have become key success factors. We have developed a usability inspection technique (WDP - Web Design Perspectives-Based Usability Evaluation) specific for Web applications' usability evaluation. The results of two previous experimental studies indicate the feasibility of the WDP technique showing it to be more effective than and as efficient as Nielsen's Heuristics Evaluation. This work describes a recent step of our research, where we examined how inspectors apply the WDP technique. In order to achieve our goal, we executed an observational study, results of which lead us to two different versions of the WDP technique.

Keywords: Web usability, Usability evaluation, Web Engineering, Web Quality.

1 Introduction

Usability is one of the three quality criteria part of the dominant development drivers for Web companies [1], and is defined by the ISO 9241 standard as "the extent to which a product can be used by specified users to achieve specified goals with effectiveness, efficiency and satisfaction in a specified context of use".

Usability is also considered a fundamental factor of Web applications' quality because of these applications' intrinsic characteristics (interactive, user-centered, hypermedia-based), where their user interface plays a vital role [2]. In addition, given that users' acceptability of Web applications seems to rely strictly on the applications' usability [3], applications with poor usability are quickly replaced by others more usable, as soon as their existence becomes known to the target audience [4].

Web Usability has two main objectives: (1) to drive the design of Web Applications and (2) to evaluate the relevant usability criteria of Web Applications. Defining methods and techniques for ensuring usability is therefore one of the goals of Web engineering researchers [3].

S. Hartmann et al. (Eds.): WISE 2008, LNCS 5176, pp. 36–47, 2008.
© Springer-Verlag Berlin Heidelberg 2008

Since 1997, with the growth in the number of Web applications, some works proposed the definition or the adaptation of usability evaluation methods to support the specific features of Web Applications [5, 6, 7, 8, 9]. This need for sounded Web usability techniques has also motivated our research goal. To this end we have proposed the Web Design Perspectives-Based Usability Evaluation (WDP) technique, based on the combination of Web design perspectives adapted from existing literature, and Nielsen's Heuristic Evaluation [10]. To support its development and validation, we adopted the experimental methodology presented in [11] (see Fig.1). This methodology extends the work of [12], and comprises six sequential steps: 1) to carry out secondary studies to identify, evaluate and interpret all available research relevant to a particular research question or topic area [13]; 2) to propose an initial version of a software technology/technique informed by the results of the secondary studies; 3) to carry out feasibility studies to assess the technology's/technique's viability; 4) to carry out observational studies to improve the understanding and the cost-effectiveness of the proposed technology/technique; 5) to carry out case studies using real development life cycles to assess the technology's/technique's suitability; and 6) to carry out case studies in industry to identify the suitability of the technology/technique in an industrial setting.

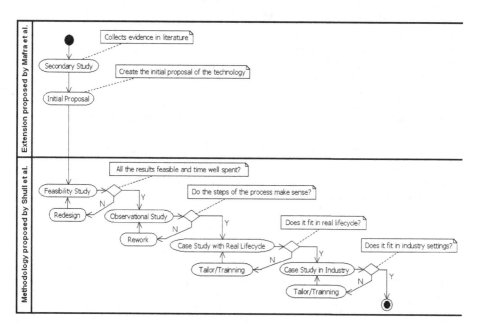

Fig. 1. Experimental Methodology Overview [14]

To date we have carried out one secondary study [15], two feasibility studies [14, 16], an observational study and a case study using a real lifecycle. The aim of this paper is to detail the observational study conducted, aimed at eliciting the most suitable process model to be used with the WDP technique; however, first we present a summary of our secondary study and both feasibility studies to provide readers with a more suitable context in which to position the observational study.

The remainder of this paper is organised as follows: Sections 2 summarises the results of the secondary study, followed by Section 3 where the first version of the WDP technique is presented. Sections 4 and 5 summarise the results of the first and second feasibility studies, respectively. Section 6 describes the Observational Study. Section 7 describes the initial WDP-RT proposal, an evolved WDP reading technique, motivated by the qualitative analysis results. Finally, conclusions and comments on future work are presented in Section 8.

2 Secondary Study: Systematic Review of Web Development Processes

The aim of our secondary study was to characterise Web Development Processes and, in addition, their quality assurance activities, more specifically inspection related activities. To this aim we conducted a systematic literature review (SR) [15, 17,18] using the guidelines outlined in [13]. The two research questions addressed in the SR were as follows:

- Q1: What development processes have been used to develop Web applications?
- Q2: What processes have been used to inspect Web applications for quality control?

The literature search retrieved 19 different Web development processes (Q1), and none of these described the use of Web quality assurance techniques such as reviews or inspections (Q2). As a result of Q1, we also identified design perspectives commonly used in Web development:

- **Conceptual:** represents the conceptual elements that make up the application domain.
- **Presentation:** represents the characteristics related to application layout and arrangement of interface elements.
- **Navigation:** represents the navigational space, defining the information access elements and their associations.
- **Structural:** represents the structural and architectural characteristics of the application, that is, how the application is structured in terms of components and their associations.

The results of the SR also prompted us to focus on a specific Web quality criterion – usability, and informed the proposal of a Web usability inspection technique, described in the next Section.

3 The Web Design Perspectives-Based Usability Evaluation – Version 1

The WDP technique prescribed the use of Usability inspection sessions to identify usability problems via the application of 13 usability heuristics (10 heuristics proposed by Nielsen [10] and three adapted from Zhang et al. [19]) in light of four separate perspectives. Each perspective corresponded to one of the four Web design perspectives that resulted from the secondary study (Conceptual (C), Presentation (P),

Navigation (N) and Structural (S)). The assumption behind the use of perspective-based inspections was that, thanks to the focus, each inspection session could detect a greater percentage of defects in comparison to other techniques that do not use perspectives. In addition, the combination of different perspectives could also detect more defects than the same number of inspection sessions using a general inspection technique [19]. Therefore, within the context of this work, Web design perspectives would be used as a guide to interpret 13 usability heuristics. Table 1 shows the associations between heuristics and Web design perspectives in WDP's version 1. In this table, the first ten heuristics represent the set proposed by Nielsen [10] and the last three correspond to the heuristics adapted from Zhang et al. [19]. The correlated pairs of heuristics-perspectives are marked with the ✓ symbol.

Hints were provided for each pair heuristic-perspective to guide the interpretation of each heuristic from a perspective's viewpoint. Note that at this stage we did not prescribe or suggest a process model to be used with the WDP technique. The WDP technique was refined by means of two consecutive feasibility studies, described in the next two Sections.

Table 1. Relationships between Heuristics and Design Perspectives in WDP v1

#H	Heuristics	Web Design Perspectives			
		C	P	N	S
1	Visibility of system status	✓	✓		✓
2	Matching between system and real world	✓			
3	User control and freedom			✓	
4	Consistency and standards		✓		
5	Error prevention	✓	✓	✓	✓
6	Recognition rather than recall	✓	✓		
7	Flexibility and efficiency of use			✓	
8	Aesthetic and minimalist project		✓		
9	Help users recognize, diagnose and recover from errors	✓	✓	✓	✓
10	Help and documentation	✓	✓	✓	✓
11	Minimize user's memory load and fatigue		✓	✓	
12	Visually functional design		✓		
13	Facilitate data entry				✓

4 First Feasibility Study

The first feasibility study was carried out in June 2006 [14]. Its goal was to compare the WDP technique to Heuristic Evaluation, using the same set of thirteen heuristics as WDP v1. We called this set of thirteen heuristics HEV+, to differ from Nielsen's set of ten heuristics, commonly used in Heuristic Evaluation (HEV). Note that, even though there are others usability evaluation techniques proposed for Web Applications, the comparison is between WDP technique and Heuristic Evaluation because WDP technique is derived from Heuristic Evaluation.

In this study we measured the number of defects found while inspecting an existing Web application. Participants were 20 undergraduate students attending a Human-Computer Interaction course at the Federal University of Rio de Janeiro (UFRJ), equally distributed in four teams, according to their experience in Software Developing (High,

Medium, or Low). Two teams applied the HEV+ technique and the other two teams applied WDPv1 Technique. Results showed that high & medium-experience participants found twice as much defects when using the WDP technique as compared to the HEV+ technique. In addition, low-experience subjects found three times as much defects when using the WDP technique as compared to the HEV+ technique. Finally, the two teams who used the WDP technique, despite differing largely on their experience levels, did not present a significant difference in the number of reported defects, suggesting that the WDP technique's effective use does not seem to depend heavily upon inspectors' expertise.

The WDP technique was revisited in light of the results from the first feasibility study, leading to the second version of the WDP technique (WDP v2). We used the quantitative results to investigate further the WDP technique and its Heuristic x Perspective (HxP) Pairs. We could notice overlaps (equivalent defects found by different inspectors) between the defects found by the HxP Pairs derived from the three heuristics from Zhang et al. [19] and defects found by the others HxP Pairs (derived from Nielsen's heuristics [10]). Due to this, the main change from version 1 to version 2 was the removal of the three heuristics adapted from Zhang et al. [19], reducing the heuristics to the set proposed by Nielsen [10].

Since we were unable to establish in the first feasibility study any relationship between the number of detected defects, the total number of existing defects and the time spent by inspectors, a second feasibility study was carried out, and summarized in the next Section.

5 Second Feasibility Study

The second feasibility study was carried out during November - December 2006 using the WDP v2 technique [16]. Its goal was to compare the efficiency and effectiveness between the WDP v2 technique and the original HEV technique. Within the context of this study efficiency and effectiveness were defined as follows:

- Efficiency: the ratio between the number of detected defects and the time spent in the inspection process.
- Effectiveness: the ratio between the number of detected defects and the total number of existing (known) defects.

Fourteen subjects participated in the study, all from different courses taught at UFRJ (three undergraduate students, nine M.Sc. students and two PhD students).

The results of the statistical analysis showed that the WDP v2 technique was significantly more effective than and as efficient as Nielsen's Heuristic Evaluation [16]. In addition, the results of this study were also used as input to further improve the WDP technique, resulting in its third version (WDP v3). The main change from the second to the third version of the WDP was the removal of the Structural Perspective due to its use being redundant.

Since the results obtained from the two feasibility studies indicated the WDP's feasibility to be more effective than, and as efficient as, the Heuristic Evaluation [10], we went one step further on following the experimental methodology, in order to also answer the second question of the methodology (see Figure 1): "Do the steps of the process make sense?". To answer this question, we carried out an observational study that had the following research goal: to elicit the process used by usability inspectors when applying the WDP technique. This study and its results are presented in the next Section, and are the main focus of this paper.

6 Eliciting the Inspection Process

Observational studies are used to understand how particular tasks are accomplished, so to obtain a fine-grained understanding of current work practices [20, 21]. To be encompassing, we gathered two types of data: observational and inquisitive. The observational data were gathered during the inspection process, without the researcher's interference [12]. As for the observational data, we used two data gathering methods: (1) the observer-executor method, in which subjects were divided in pairs with two roles: the "executor", who carries out the inspection and the "observer", who watches carefully how the "executor" conducts the inspection; (2) the cooperative evaluation, which is a "Think aloud" method variation[21]. The cooperative evaluation was the interaction protocol used by each pair observer-executor, because the executor describes (or "thinks aloud") what (s)he is doing and the observer is free to ask questions/explanations about the executor's decisions or acts [21]. The inspection was divided in two parts such that all subjects would be able to play both roles, i.e., the subjects that were "observers" in Part I, became "executors" in Part II and vice-versa. Inquisitive data was gathered at the completion of each process step, rather than during its execution, using follow-up questionnaires.

According to the experimental methodology described in [11, 12], the goal of the observational study should be to answer the question: "Do the steps of the process make sense?". However, the WDP technique is a checklist-based inspection technique, and as such does not have an explicit order of steps to be followed. According to Zhang et al. [19], we know that focus at one perspective at a time would bring better inspection results. Then, we recommended that the inspectors applied the WDP (v3) technique focusing at one perspective at a time. And we also suggested the sequence in which to apply the perspectives: first the heuristics related to the Presentation Perspective, followed by the heuristics related to the Navigational Perspective and finally the heuristics related to the Conceptual Perspective. This order was chosen based on the results of the second feasibility study for each perspective [14]. Note that: (1) we only suggested a sequence to be used with the Perspectives instead of a complete process to follow; and (2) the sequence that was suggested was only a recommendation, so subjects could still decide about using a different sequence when applying the Perspectives.

The planning and execution of this observational study is detailed in Section 6.1. Herein our aim was to understand deeply the WDP process, so we did not compare the WDP with any other technique.

6.1 Observational Study

We performed this observational study in the 1st semester of 2007, with the participation of 14 undergraduate students attending a HCI course offered at UFRJ. Prior to participating in this study, subjects were trained in usability techniques, such as heuristic evaluation [10] and cognitive walkthrough [22]. During the study, they were trained in the WDP technique, the Web application to inspect, and the Observer-Executor method. Subjects were split into two groups (A and B), each containing seven subjects, which were arranged using as criterion subjects' assignments grades. Group A contained the top seven students and Group B the remaining students. All the Observer-Executor pairs had one student from each group. The Web application used was the same one from the feasibility studies [14, 16], and the same inspection domain, the six use cases associated with the role 'Conference Reviewer'. Part I was carried out first and inspected the four simplest use cases. Later, Part II inspected the two remaining use cases, which were the most complex.

During Part I, Group B subjects were the "executors" and Group A the "observers". As previously mentioned, "executors" were asked to apply the WDP technique in a set sequence. Once Part I was completed, "executors" provided a worksheet with the defects found and a follow-up questionnaire containing their impressions regarding the WDP technique; "observers" provided forms containing any notes taken during Part I. During Part II, Group A subjects now became the "executors". Notice that, at this point, Group A subjects had already a good grasp of usability techniques (reflected in their grades), and in addition had also observed Group B applying the WDP technique, which in our view increased their knowledge of usability inspection and the WDP technique. We stressed to the participants that if they did not agree with to use the suggested application sequence, they could use any other sequence as they saw fit. Finally, we also observed carefully how this group chose to apply the WDP technique.

The quantitative data were gathered via the defects' worksheets, used to identify real defects and false positives. Results showed that Group A subjects found a very small number of false positive defects in Part II, thus suggesting they were experienced in usability inspection. The qualitative data were gathered by means of the follow-up questionnaires and the observational forms, and analyzed using concepts from Grounded Theory [23], as detailed in Section 6.2.

6.2 Qualitative Data Analysis

The qualitative data were analyzed using a subset of the steps part of the Grounded Theory (GT) method [23] – the open/axial coding activities. This method is a qualitative method for data analysis, where data systematically gathered and analyzed through the research process is used to derive a theory [23]. It is based on coding – the analytic processes through which data are fractured, conceptualized, and integrated to form a theory, and contains three data analysis steps: open coding, where concepts are identified and their properties and dimensions are discovered in data; axial coding, where connections between the categories (and sub-categories) are identified; and selective coding, where the core category (that integrates the theory) is identified and described [23]. The process of grouping concepts that seem to pertain to the same phenomena is called categorizing, and it is done to reduce the number of units to work with [23]. In this observational study it was not necessary to execute all three data analysis steps, because we could get the answer to our research question ("how do the inspectors apply WDP technique?") after the execution of open/axial coding activities. For this reason, we do not claim that we applied the GT method, only some specific procedures.

The data analysis began with the open coding applied the follow-up questionnaires. The objective of the open coding activity was to analyze the data collected and allocate codes to the text. We did not use "seed categories" (an initial set of codes), but rather coded directly from text, creating in-vivo codes. The open coding procedures stimulate the constant creation of new codes and the merging of existing codes as new evidence and interpretations emerged. The open coding of all 14 questionnaires produced 74 in-vivo codes linked to 162 quotations within the questionnaires. In the next step - axial coding, codes are grouped according to their properties, thus forming concepts that represent categories. These categories are analyzed and subcategories are identified aiming to provide more clarification and specification. Finally, the categories and subcategories are related to each other, and the causal relationships between categories are determined.

In practice, the open and axial coding steps overlapped and merged because the process proceeded iteratively. At the end of this analysis, the coding processes produced altogether 93 codes (with 74 in-vivo codes), classified into 04 main categories. Figure 2 shows one of the four categories (Problem category) with the related codes and relationships.

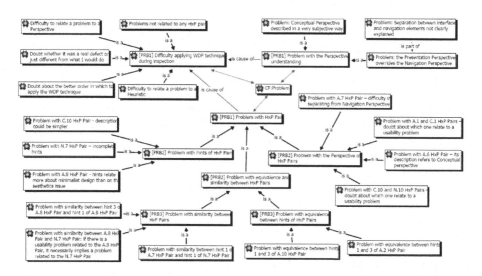

Fig. 2. The relationships between the codes of one category

Seaman [20] claims that qualitative data can be used to go beyond the statistics and help to explain the reasons behind the hypotheses and relationships. When we use analytic methods to examine qualitative data, we achieve a much deeper understanding of the whole phenomena. The systematic browsing through the data and coding the occurrences of the WDP process-related phenomena enables one to understand the WDP learning process of a usability inspector. And we could notice that there were large differences between the WDP learning process of a novice inspector and the

WDP learning process of an inspector with more expertise (or more training) in usability evaluation.

Whenever inspectors did not have a good usability evaluation knowhow, they seemed to prefer further directions on how to apply the WDP technique. According to the reported questionnaires, when a novice inspector started to evaluate a web application interaction, s(he) felt insecure about how to start, because each perspective has many pairs HxP to apply. The novice inspector wants to know what (s)he has to look for, a kind of guiding procedure as: "first look for X, then verify if X is…". Being a checklist-based technique, WDP does not have this type of detailed procedural steps.

Some inspectors reported that, as they did not know any prescribed sequence in which to apply the HxP pairs, they simply did not follow any order at all when applying the WDP technique. They just browsed the Web application, looking for the usability problems. When they identified a possible usability problem, they tried to relate the problem to a specific HxP pair. But many times they felt unsure as to which Heuristic and/or Perspective would be related to the usability problem identified.

While identifying the variety of conditions, interactions and consequences associated with this process when applying WDP technique, we came across to important issues that showed us that we should develop a different type of WDP technique, aimed at reducing novice inspectors' difficulties. This "novice-focused" WDP technique should provide more guidance and reduce the cognitive effort in applying the technique.

According to the qualitative analysis, an inspector with good knowhow in usability begins applying the WDP technique following the Design Perspectives focus – for each interaction, they tend to apply first the presentation perspective, followed by conceptual perspective and finally the navigation perspective, which is only used when (s)he navigates through the interaction. Some of the experienced inspectors reported that they grasped the WDP technique principles after inspecting the application for a short period, suggesting that the WDP's learning curve was small. Once experienced inspectors fully grasped the technique, they ruled out the need to follow any prescribed order in which to apply the WDP technique. As described in [12], inspectors tend to adjust the technique to their own way of thinking about a problem.

Therefore we noticed that for inspectors with good knowhow of usability evaluation, an inspection process with a fixed sequence of steps was not necessary in order to apply the technique, and the checklist-based WDP version was adequate to their needs. However, the use of a checklist-based inspection did not seem the best approach for novice inspectors. Whenever inspectors did not have a good usability knowhow, they seemed to prefer further directions on how to apply the WDP technique, as they felt unsure, when identifying a usability problem, to which Heuristic and/or Perspective that problems related to.

The general rule in grounded theory is to sample until a theoretical saturation is reached. This means that sampling should continue until: (1) no new or relevant data seem to emerge regarding a category; (2) the category development is dense, insofar as all of the paradigm elements are accounted for, along with variation and process; (3) the relationships between categories are well established and validated [23]. In our research, the theoretical saturation was not yet reached, because we had had only one data collection round. Recently we conducted another study to be used to complement the one presented herein; however its data are yet to be analyzed.

7 Initial Proposal for an Evolved Usability Inspection Technique

The analysis of the qualitative data provided us with important feedback to improve further the checklist-based WDP version, as it pointed out specific problems in some HxP pairs. These problems were analyzed by three usability researchers and this led to a new version of the WDP technique (WDP v4), with improvements in some hints of HxP pairs. The qualitative results from the observational study were also used to extend the WDP technique into a reading technique called WDP-RT. This allows usability inspectors to choose between two categories of usability inspection techniques: (1) WDP v4 – the checklist-based technique, with its HxP pairs improved with the results of the observational study; and (2) WDP RT – the evolved reading technique, aimed at reducing novice inspectors' difficulties.

According to Travassos et al. [24], a reading technique is a specific type of inspection technique that has a series of steps for the individual analysis of a software product to achieve the understanding needed for a particular task. Reading techniques attempt to increase the effectiveness of inspections by providing procedural guidelines that can be used by individual reviewers to examine a given software artifact and identify defects [24]. We believe that, having those procedural guidelines, it would be easier for a novice inspector to execute a usability evaluation.

A reading technique provides focus, which is used to guide an inspector during a review activity. In the case of the WDP RT, the focus is on each one of the Design Perspectives. The WDP-RT will present one perspective at a time, showing the definition and an abstraction of the perspective, in order to support an inspector to fully understand the perspective. Next, the WDP-RT will list the Usability Features related to that focus. The Usability Features will present all the issues that should be verified in a given perspective. A Usability Feature represents an abstraction of a Heuristic, a type of component that incorporates and couples heuristic(s), so reducing inspector's effort when correlating problems with their corresponding heuristics.

Our proposal is to apply the set of "Functional Usability Features" suggested by Juristo et al. [25] to Nielsen's Heuristics and the Perspectives part of the WDP technique. The motivation to consider the set of Functional Usability Features, in addition to the Nielsen's Heuristics, is that it represents the result of a vast research about usability features with relevant usability benefits, according to the usability literature. Besides, they were empirically evaluated [25]. Therefore, WDP RT aims to provide close guidance to usability inspectors, by offering a procedure (and a suggested order) to review all of the Usability Features for each design perspective.

Finally, experienced inspectors, for whom detailed guidance is not necessary, can continue to use the checklist-based WDP (now v4). The advantage of using a checklist-based inspection technique is that it highlights the important points that should be reviewed, and the inspectors can be free to apply the technique as they see fit.

8 Conclusions and Comments on Future Work

This paper has provided an overview of the studies carried out to date to propose a Web usability inspection technique, with particular emphasis on the observational study. This study aimed at eliciting the process used by usability inspectors when applying the checklist-based WDP technique.

The results of the observational study showed that there was no single most effective and efficient process to be used with the checklist-based WDP; and there were large differences between the WDP learning process of a novice inspector and the WDP learning process of an inspector with more knowhow in usability evaluation. This led us to propose a new different type of WDP technique, a reading technique called WDP-RT.

Although the results of a single experience cannot be generalized to other contexts, we believe that the qualitative results of this observational study about the differences between the learning process of a novice and a non-novice inspector could contribute to improve the general understanding about inspections.

Future work entails: (1) the analysis of the qualitative data from the case study, (2) a case study in industry to further validate the checklist-based WDP, carrying out the series of empirical studies suggested by the experimental methodology [11], (3) a feasibility study to validate the new WDP-RT technique and, (4) the development of automated support for both techniques such that their acceptance in industry is facilitated.

Acknowledgments. The authors would like to thank C. R. de Souza and M. Montoni for their important help with GT. We would also like to thank V. Bravo for his assistance during the observational study, and all the students who participated in the studies. This work has been partially supported by the ESE and Science in Large Scale Project CNPq (475459/2007-5), FAPERJ and FAPEAM.

References

1. Offutt, J.: Quality attributes of Web software applications. IEEE Software 19(2), 25–32 (2002)
2. Olsina, L., Covella, G., Rossi, G.: Web Quality. In: Mendes, E., Mosley, N. (eds.) Web Engineering, pp. 109–142. Springer, Heidelberg (2006)
3. Matera, M., Rizzo, F., Carughi, G.: Web Usability: Principles and Evaluation Methods. In: Mendes, E., Mosley, N. (eds.) Web Engineering, pp. 143–179. Springer, Heidelberg (2006)
4. Mendes, E., Mosley, N., Counsell, S.: The Need for Web Engineering: An Introduction. In: Mendes, E., Mosley, N. (eds.) Web Engineering, pp. 1–26. Springer, Heidelberg (2006)
5. Instone, K.: Site Usability Evaluation. In: Web Review (October 1997)
6. Nielsen, J.: Designing Web Usability. New Riders (1999)
7. Brinck, T., Gergle, D., Wood, S.: Usability for the Web: Designing Web Sites that Work. Morgan Kaufmann, San Francisco (2002)
8. Matera, M., Costable, M., Garzotto, F., Paolini, P.: SUE Inspection: An Effective Method for Systematic Usability Evaluation of Hypermedia. IEEE Transactions on Systems, Men, and Cybernetics 32(1) (January 2002)
9. Triacca, L., Bolchini, D., Botturi, L., Inversini, A.: MiLE: Systematic Usability Evaluation for E-learning Web Applications. In: Cantoni, L., McLoughlin, C. (eds.) Proc. ED-MEDIA 2004 World Conf.on Educational Multimedia, Hypermedia & Telecommunications (2004)
10. Nielsen, J.: Heuristic evaluation. In: Nielsen, J., Mack, R.L. (eds.) Usability Inspection Methods, pp. 25–62. John Wiley & Sons, Chichester (1994)

11. Mafra, S.N., Barcelos, R.F., Travassos, G.H.: Aplicando uma Metodologia Baseada em Evidência na Definição de Novas Tecnologias de Software. In: Proc. of the 20th SBES 2006, Portuguese, Florianópolis, Brazil, pp. 239–254 (2006)
12. Shull, F., Carver, J., Travassos, G.: An Empirical Methodology for Introducing Software Processes. ACM SIGSOFT Software Engineering Notes 26(5), 288–296 (2001)
13. Kitchenham, B.: Procedures for Performing Systematic Reviews. Joint Technical Report, Keele University, UK, TR/SE-0401 and NICTA Technical Report 0400011T.1 (2004)
14. Conte, T., Massolar, J., Mendes, E., Travassos, G.: Web Usability Inspection Technique Based on Design Perspectives. In: Proc. of the SBC 21th Brazilian Symposium on Software Engineering (SBES 2007), João Pessoa, Brazil (2007)
15. Conte, T., Mendes, E., Travassos, G.: Development Processes for Web Applications: A Systematic Review. In: Proc. of the SBC 11th Brazilian Symposium on Multimedia and Web (WebMedia 2005), Portuguese Poços de Caldas, Brazil (2005)
16. Conte, T., Massolar, J., Mendes, E., Travassos, G.: Usability Evaluation Based on Web Design Perspective. In: First International Symposium on Empirical Software Engineering and Measurement (ESEM 2007), Madrid, Spain, pp. 146–155 (2007)
17. Mendes, E. and Kitchenham, B.: Protocol for Systematic Review (2005), http://www.cs.auckland.ac.nz/emilia/srspp.pdf
18. Mian, P., Conte, T., Natali, A., Biolchini, J., Mendes, E., Travassos, G.: Lessons Learned on Applying Systematic Reviews to Software Engineering. In: 3rd Intern. Workshop GUIDELINES FOR EMPIRICAL WORK, WSESE 2005, Finland (2005)
19. Zhang, Z., Basili, V., Shneiderman, B.: Perspective-based Usability Inspection: An Empirical Validation of Efficacy. Empirical Software Engineering: An International Journal 4(1), 43–69 (1999)
20. Seaman, C.: Qualitative Methods in Empirical Studies of Software Engineering. IEEE Transactions on Software Engineering 25(4) (1999)
21. Dix, A., Finlay, J., Abowd, G., Beale, R. (eds.): Human-Computer Interaction, 3rd edn. Prentice Hall, Englewood Cliffs (2004)
22. Polson, P., Lewis, C., Rieman, J., Wharton, C.: Cognitive Walkthroughs: a method for theory-based evaluation of users interfaces. Int. Journal of Man-Machine Studies (1992)
23. Strauss, A., Corbin, J.: Basics of Qualitative Research: Techniques and Procedures for Developing Grounded Theory, 2nd edn. SAGE Publications, Thousand Oaks (1998)
24. Travassos, G., Shull, F., Carver, J.: Reading Techniques for OO Design Inspections. Technical Report ES-575/02, PESC/COPPE/UFRJ (2002)
25. Juristo, N., Moreno, A., Sanchez-Segura, M.: Guidelines for Eliciting Usability Functionalities. IEEE Transactions on Software Engineering 33(11), 744–758 (2007)

Ranking People Based on
Metadata Analysis of Search Results

Qiang Ma[1] and Masatoshi Yoshikawa[1]

Graduate School of Informatics, Kyoto University,
Yoshida Honmachi, Sakyo, Kyoto, 606-8501, Japan
{qiang,yoshikawa}@i.kyoto-u.ac.jp
http://www.db.soc.i.kyoto-u.ac.jp

Abstract. Person search is one of the most popular search types on the Web. Most of the conventional technologies for person search focused on mapping the person name to a specific person (i.e. referents). In contrast, in this paper, we propose a novel ranking measure called famousness for person search. We use the notion of famousness for ranking people according to how well-known they are. Intuitively, famousness score is computed by analyzing the metadata of search results returned by a search engine. The metadata used in our method include URL, snippet, and the number of search results. To compute the famousness score of a person, first, we cluster the search results by using their metadata. Second, we compute the deviations in the size and number of such clusters. If the related Web pages of a person can be grouped into many large clusters of similar size, it looks like that person has been mentioned in many Web pages from various domains and that s/he is well known. In addition, we compare the clusters of search results with those of other people. Persons having more and larger clusters are given higher famousness scores. We also show experimental results to validate the ranking method based on the famousness score.

1 Introduction

The Web reflects the real world and at the same time is bringing great changes to our daily lives. For instance, information related to people is available on the Web, and it impacts on our daily activities. For example, usually, there are many Web pages related to a university professor and these Web pages are very useful when a student looks for a supervisor candidate when applying to graduate school.

To support users in finding information related to certain people, person search has come into the spotlight. Most of the conventional person search methods focus on how to identify the reference relationship between Web pages and persons[4, 6, 10]. On the other hand, there are also systems that automatically generate knowledge-bases from the Web. One example is the YAGO[9], which automatically builds a knowledge-base from the free encyclopedia Wikipedia and WordNet. By using a knowledge base automatically or semi-automatically

S. Hartmann et al. (Eds.): WISE 2008, LNCS 5176, pp. 48–60, 2008.

built from the Web, it is possible to make systems for supporting knowledge searches[2, 5]. These knowledge based systems make it easy to find people who are very famous now or in history.

When we submit a person query, these conventional technologies often return us more than one person as the results. To the best of our knowledge, though, there is no system which supports ranking. Recall the example in which a student wants to find information on professors when s/he applies to graduate school. Suppose that s/he has a list of professors in the subject field of database area or a list of professors in the computer science department of K-university; s/he may want to shortlist the candidates and rank them. In this paper, we propose a solution to assist her/him in such task.

We propose a novel method to rank people by using the metadata of search results returned by a search engine. The metadata used in our method include URL, snippet, and the number of search results. Although a mapping between the Web pages and a certain person is necessary in our method, it is not the major contribution of this paper. We focus rather on the next step of the person search: that is, to rank persons in the real world by using the information available on the Web.

In this paper, we propose a notion called famousness as one of the ranking measures for person search. We use famousness score to rank people according to how well-known they are. Web page related to a person could be looked as one kind of reputations for her/him. Hence, the related Web pages of a person could be used to estimate how well-known s/he is. The simplest way is to rank people by using the number of their related Web pages. However, in actuality, someone, who belongs to a big community or has many related people (students, etc.), may have a large number of Web pages that are mostly hosted on her/his official and related Web sites. S/he may be famous in some special organizations and communities rather than well-known. Therefore, it is important to consider well the variations of their related Web pages to rank people rather than the number of pages. In this paper, we compute the famousness scores to rank people based on variation analysis of Web pages in three ways: content by analyzing snippets, host of Web site by analyzing URL, and organization by analyzing domain type.

Currently, the famousness score of a person is computed in two steps.

- Inside Analysis: first, we analyze the metadata of results returned by a search engine per person. We cluster the search results by using metadata based on snippet similarity, URL similarity, and domain type, respectively. As the results, there are three kinds of clusters: content clusters, URL clusters, and type clusters. Then, we compute the deviation of each kind of clusters and with these values compute the inside score of famousness. The basic idea is that if we can group the search results into many large clusters of similar size, that person may have been mentioned in many Web pages from various domains and the probability that s/he is well-known is high.

- Comparison Analysis (or Outside Analysis): we also compare the clusters of the search results with those of other people. People having more and bigger clusters than others are given higher famousness scores.

One of the notable features of our method is that we utilize the metadata of search results to rank people in a quick way and further information is not necessary. When the user enters a person list, our method generates person queries and submits them to a conventional search engine. The results and their metadata of each person are screened to exclude junk pages not containing personal information and pages related to other persons having the same name. After that, we compute the famousness score to rank these people.

The remainder of this paper is organized as follows. In Section 2, we introduce related work. In Section 3, we describe the notion of famousness. We show experimental results in Section 4. We summarize the paper and look at future research in Section 5.

2   Related Work

Person search is one of the most popular search types on the Web. Most of the conventional person search methods focus on how to identify the reference relationship between Web pages and persons.

Clustering approaches are widely used in the conventional person search methods. For instance, the person resolution system WebHawk[10] tries to group search results by using the features of person name, properties extracted from Web pages, and so on. It implements functions of removing junk pages and mapping between page and person. WebHawk groups person pages into different clusters by using an agglomerative clustering algorithm. One of its notable features is that it considers well full English person names, which are usually composed of three fields: first name, middle name, and last name. However, it's difficult or inappropriate to apply this feature into person queries in other languages.

Al-Kamaha and Embley[4] proposed a mapping method that is based on clustering search results with the person attributes, links, and page similarity. Mann and Yarowsky[6] extract the biographic features to generate person clusters based on a bottom-up centroid agglomerative clustering algorithm. Instead of clustering, Guha and Garg[3] use a re-ranking algorithm to disambiguate people in an interactive manner. Methods using social networks are also proposed[1]. However, to the best of our knowledge, there is no method proposed to rank people.

Knowledge-base is a kind of useful resource for finding famous people. Systems that automatically generate knowledge bases from the Web have been investigated. One example is the YAGO[9], which automatically builds a knowledge-base from the free encyclopedia Wikipedia and WordNet. Famous persons, especially famous people in history, are stored in the knowledge base as entities or facts for further use. By using a knowledge-base automatically or semi-automatically built from the Web, it is possible to make systems for supporting knowledge searches[2, 5]. These systems help us to find people who are very famous now or in history. However, the results are limited to a very narrow scope of famous people and there is no ranking technique.

3   Ranking People Based on Famousness Score

As mentioned above, famousness is a notion computed from a metadata analysis of search results for ranking people. We compute the famousness score of a person in two ways: 1) by analyzing metadata of search results returned for that person to compute an inside score of famousness, and 2) by comparing search results returned for that person and the others to compute an outside score of famousness. The final integrated score of famousness is computed from the inside and outside scores.

```
<ResultSet xmlns:xsi="http://www.w3.org/2001/XMLSchema-instance"
xmlns="urn:yahoo:srch" xsi:schemaLocation="urn:yahoo:srch
http://search.yahooapis.com/WebSearchService/V1/WebSearchResponse.xsd"
totalResultsAvailable="3610652" totalResultsReturned="1"
firstResultPosition="1">
 <Result>
  <Title>Madonna</Title>
  <Summary>official site, with news, music, media, and fan club. Includes
information on Madonna's 2004 Re-Invention tour.</Summary>
   <Url>http://www.madonna.com/</Url>
<ClickUrl>http://rds.yahoo.com/S=2766679/K=madonna/v=2/XP=yws/SID=e/l
=WS1/R=1/SS=33338411/H=1/IPC=us/SHE=0/SIG=11b8pqr3e/EXP=1106183
712/*-http%3A//www.madonna.com/</ClickUrl>
  </Result>
 </ResultSet>
```

Fig. 1. Example of Metadata of Search Results Returned by Yahoo! Search Service

3.1   Metadata of Search Results

Search engines such as Google and Yahoo! provide search results metadata via their API services. For instance, the Web search service of the Yahoo! developer network [11], returns the following fields as metadata when we submit a query to it.

- *ResultSet* which contains all the query responses. The number of query matches in the database and number of returned results are also returned as attributes of ResultSet.
- *Result* which contains individual responses.
- *Title* of the Web page.
- *Summary* of the Web page. In this paper, we call this a snippet.
- *Url* for the Web page.
- *ClickUrl* which is the URL for linking to the Web page.
- *MimeType* which is the MIME type of the Web page.
- *ModificationDate* which gives the last modification date of the Web page.
- *Cache* which includes the cached result's URL and its size in bytes.

Figure 1 illustrates metadata returned for the query "Madonna" by the search service of Yahoo!.

Though their names may be different, the other search API services also provide most of these fields. The famousness score is computed by analyzing metadata including number of query matches, snippets and URLs of searched Web pages.

3.2   Inside Score of Famousness

The inside score of famousness is computed by analyzing the search results returned for the query using that person's name. For a given person name, we submit a person query to a search engine and get the search results. Then we use that person's affiliation information to remove pages not related to him/her. After the screening step, only the Web pages related to the given person will be left for computing his/her famousness score.

The related Web pages are grouped into clusters, and the deviation of these clusters' size is computed. Large number of clusters, big cluster size and small deviation of cluster size lead a larger inside score of famousness. In our current work, we group the search results according to the snippet similarity, URL similarity, and domain type, respectively.

Let p stand for the given person and the related Web pages be grouped into content clusters $C(p)$ based on snippet similarity, URL clusters $U(p)$ based on URL similarity, and type clusters $T(p)$ based on domain type, respectively. The inside scores of p's famousness are computed per kind of cluster, as follows.

$$famous_i(p, C) = n_c \cdot \frac{s_c}{\sigma_{s_c}} \tag{1}$$

$$famous_i(p, U) = n_u \cdot \frac{s_u}{\sigma_{s_u}} \tag{2}$$

$$famous_i(p, T) = n_t \cdot \frac{s_t}{\sigma_{s_t}} \tag{3}$$

where $n_x (= |X(p)|)$ denotes the number of clusters of $X(p)$. s_x and σ_{s_x} respectively denote the mean and the standard deviation of the cluster size of $X(p)$.

In the rest of this subsection, we will respectively explain the clustering methods based on snippet similarity, URL similarity and domain type.

Clustering Based on Snippet Similarity. The Web pages related to the given person are grouped into clusters by using the complete linkage method. The similarity between pages is computed using the keyword vectors of their snippets as follows.

$$sim(s_1, s_2) = \frac{k_{11} \cdot k_{21} + k_{12} \cdot k_{22} + ... + k_{1n} \cdot k_{2n}}{\sqrt{k_{11}^2 + ... + k_{1n}^2} \cdot \sqrt{k_{21}^2 + ... + k_{2n}^2}} \tag{4}$$

where s_1 and s_2 stand for the snippets of Web pages, respectively. $(k_{11}, k_{12}, ..., k_{1n})$ and $(k_{21}, k_{22}, ..., k_{2n})$ are the $tf \cdot idf$ keyword-vectors of s_1 and s_2, respectively. Here, the idf value is computed within the related Web pages.

The number of clusters, the mean and the standard deviation of cluster size are computed for later computation of the famousness score. Suppose that the cluster number is $n_c(= |C(p)|)$; the mean and the standard deviation of cluster size are computed as follows.

$$s_c = \frac{1}{n_c} \Sigma_{i=1}^{i=n_c} |c_i| \tag{5}$$

$$\sigma_{s_c} = \sqrt{\frac{1}{n_c - 1} \Sigma_{i=1}^{i=n_c} (|c_i| - s_c)^2} \tag{6}$$

where $|c_i|$ stands for the size (the number of related pages) of cluster c_i.

Table 1. Domain Types for Clustering

Domain Type	TLDN	SLD
Education	"edu"	"ac.jp", "ed.jp"
JPNIC members	-	"ad.jp"
Company	"com"	"co.jp"
Government	"gov"	"go.jp"
Local	-	"lg.jp"
Network	"net"	"ne.jp"
Org	"org"	"or.jp"
Other	other	other

Clustering Based on Url Similarity. Intuitively, URL similarity denotes the probability of two web pages being hosted on the same site. Based on URL similarity, we group the related Web pages of a person into Web sites in order to analyze the site variation of related Web pages.

We also use the complete linkage method to group Web sites by using URL similarity. The URL similarity of two URLs url_1 and url_2 is computed as follows.

1. Extract the host name h_1 and h_2 from url_1 and url_2, respectively. For example, the host name of "http://www.i.kyoto-u.ac.jp/school/index.html" is "www.i.kyoto-u.ac.jp".
2. Split h_1 and h_2 into sub-string arrays S_1 and S_2, respectively. The split key is ".". The sub-strings of the host name are stored in reverse order. For example, "www.i.kyoto-u.ac.jp" is stored in the order of "jp", "ac", "kyoto-u", "i" and "www".
3. for $i = 0$ to $min(S_1.length, S_2.length)$ do
   - if $S_1[i] == S_2[i]$, then $sim = sim + 1$;
4. $sim(url_1, url_2) = \frac{sim}{(min(S_1.length, S_2.length))}$.

The mean and standard deviation of cluster size are computed as follows.

$$s_u = \frac{1}{n_u} \Sigma_{i=1}^{i=n_u} |u_i| \tag{7}$$

$$\sigma_{s_u} = \sqrt{\frac{1}{n_u - 1} \Sigma_{i=1}^{i=n_u} (|u_i| - s_u)^2} \tag{8}$$

where $|u_i|$ stands for the size (the number of related pages) of cluster u_i. n_u $(= |U(p)|)$ is the number of clusters grouped based on URL similarity.

Clustering Based on Domain Type. The variation of domain type is an important factor to compute a person's famousness score. For example, a well known professor may have Web pages from various domains, such as the education domain, government domain, and company domain.

We group the related Web pages into clusters by using the top-level domain name (TLDN). TLDN is the last part of an Internet domain name; that is, the letters that follow the final dot of any domain name. For example, in the domain name, "www.google.com", the top-level domain name is "com". Because we focus on Web pages in Japanese, we also analyze the second-level domain (SLD) in the internet country code top-level domain (ccTLD) for Japan (".jp"). For example, in the domain name, "www.kyoto-u.ac.jp", the second-level domain is "ac.jp".

In short, to group Web pages, if the host name of a URL contains ccTLD, we use the second-level domain; if there is no ccTLD, only the TLDN will be used. We use eight domain types in Table 1 for clustering related Web pages based on the domain type.

We compute the mean and standard deviation of cluster size as follows.

$$s_t = \frac{1}{n_t} \Sigma_{i=1}^{i=n_t} |t_i| \tag{9}$$

$$\sigma_{t_s} = \sqrt{\frac{1}{n_t - 1} \Sigma_{i=1}^{i=n_t} (|t_i| - s_t)^2} \tag{10}$$

where, $|t_i|$ is the size of cluster t_i; $n_t (= |T(p)| \leq 8)$ is the number of clusters grouped based on domain type.

3.3   Outside Score of Famousness

The outside score of famousness is computed by comparing the search results returned for that person with other ones. Persons having more and larger clusters will get higher outside scores than persons having fewer and smaller clusters.

For a given person list, first, we compute the inside score of famousness of each person. In this step, we compute the standard deviation and mean of size of three kinds of clusters (content clusters, URL clusters and type clusters). Next, we compute the standard deviation of cluster size and the mean of cluster number among all persons per kind of cluster. Finally, we compute the outside scores of famousness per person. Persons whose related Web pages have been grouped into more and larger clusters than average will be given higher outside scores of famousness.

Suppose that the given person list is $P = \{p_1, p_2, ..., p_m\}$. The standard deviation and mean of clusters number among all given persons are computed per kind of cluster, as follows.

$$\overline{n_c} = \frac{1}{m} \Sigma_{i=1}^{i=m} n_{c_i} \tag{11}$$

$$\sigma_{\overline{n_c}} = \sqrt{\frac{1}{m-1} \Sigma_{i=1}^{i=m} (n_{c_i} - \overline{n_c})^2} \tag{12}$$

$$\overline{n_u} = \frac{1}{m} \Sigma_{i=1}^{i=m} n_{u_i} \tag{13}$$

$$\sigma_{\overline{n_u}} = \sqrt{\frac{1}{m-1} \Sigma_{i=1}^{i=m} (n_{u_i} - \overline{n_u})^2} \tag{14}$$

$$\overline{n_t} = \frac{1}{m} \Sigma_{i=1}^{i=m} n_{t_i} \tag{15}$$

$$\sigma_{\overline{n_t}} = \sqrt{\frac{1}{m-1} \Sigma_{i=1}^{i=m} (n_{t_i} - \overline{n_t})^2} \tag{16}$$

where n_{c_i}, n_{u_i}, and n_{t_i} stand for the numbers of content clusters, URL clusters, and type clusters of person p_i, respectively. $\overline{n_c}, \overline{n_u}$ and $\overline{n_t}$ denote the mean numbers of the three kinds of clusters. $\sigma_{\overline{n_c}}, \sigma_{\overline{n_u}}$ and $\sigma_{\overline{n_t}}$ are the standard deviations of the numbers of these three kinds of clusters.

Similarly, the standard deviation and mean of cluster size among all given persons are computed as follows.

$$\overline{s_c} = \frac{1}{m} \Sigma_{i=1}^{i=m} s_{c_i} \tag{17}$$

$$\sigma_{\overline{s_c}} = \sqrt{\frac{1}{m-1} \Sigma_{i=1}^{i=m} (s_{c_i} - \overline{s_c})^2} \tag{18}$$

$$\overline{s_u} = \frac{1}{m} \Sigma_{i=1}^{i=m} s_{u_i} \tag{19}$$

$$\sigma_{\overline{s_u}} = \sqrt{\frac{1}{m-1} \Sigma_{i=1}^{i=m} (s_{u_i} - \overline{s_u})^2} \tag{20}$$

$$\overline{s_t} = \frac{1}{m} \Sigma_{i=1}^{i=m} s_{t_i} \tag{21}$$

$$\sigma_{\overline{s_t}} = \sqrt{\frac{1}{m-1} \Sigma_{i=1}^{i=m} (s_{t_i} - \overline{s_t})^2} \tag{22}$$

where s_{c_i}, s_{u_i}, and s_{t_i} stand for the mean size of content, URL and type clusters of person p_i, respectively. $\overline{s_c}, \overline{s_u}$, and $\overline{s_t}$ denote the mean sizes of the three kinds of clusters among all given persons. $\sigma_{\overline{s_c}}, \sigma_{\overline{s_u}}$ and $\sigma_{\overline{s_t}}$ are the standard deviations of the sizes of these three kinds of clusters.

The outside scores of person p_i are computed per kind of cluster, as follows.

$$famous_o(p_i, C) = e^{\frac{n_{c_i} - \overline{n_c}}{\sigma_{\overline{c_n}}} + \frac{s_{c_i} - \overline{s_c}}{\sigma_{\overline{s_c}}}} \tag{23}$$

$$famous_o(p_i, U) = e^{\frac{n_{u_i} - \overline{n_u}}{\sigma_{\overline{u_n}}} + \frac{s_{u_i} - \overline{s_u}}{\sigma_{\overline{s_u}}}} \tag{24}$$

$$famous_o(p_i, T) = e^{\frac{n_{t_i} - \overline{n_t}}{\sigma_{\overline{t_n}}} + \frac{s_{t_i} - \overline{s_t}}{\sigma_{\overline{s_t}}}} . \tag{25}$$

3.4   Ranking People Based on Integrated Score of Famousness

For a given person list P, we compute the famousness score of person $p_i \in P$ in an integrated form of the outside and inside scores. These people in P are ranked by the integrated score of famousness. In contrast, it is also considerable that use only the inside score or the outside score to rank people, too.

Currently, we compute the integrated score of famousness as follows.

$$\begin{aligned} famous(p_i) = {} & \alpha \cdot famous_i(p_i, C) * famous_o(p_i, C) \\ & + \beta \cdot famous_i(p_i, U) * famous_o(p_i, U) \\ & + \gamma \cdot famous_i(p_i, T) * famous_o(p_i, T) \end{aligned} \quad (26)$$

where α, β, γ are weight parameters.

4 Experimental Evaluation

We carried out an experiment to validate our ranking method. Seven users including five graduate students and two undergraduate students were asked to rank five professors in their university (Prof. in K. Univ.) and five professors in the research field of database (Prof. in DB Area), respectively. Surveys using the Internet and social-network were allowed.

We implemented our ranking method by using the Yahoo! Japan Web search API[12] and Slothlib[7]. For each person, we analyzed the top 1000 pages (maximum) returned by the Yahoo! Japan Web search service. Stemming and removal of stop words were done in advance. Based on preliminary experiments, the similarity thresholds of clustering based on snippet similarity and URL similarity were 0.1 and 0.3, respectively. The parameters (α, β and γ) to compute the integrated famousness score were set to $(1, 1, 1)$ for using all cluster types, $(1, 0, 0)$ for using content cluster based on snippet similarity, $(0, 1, 0)$ for using url clusters based on url similarity, and $(0, 0, 1)$ for using type clusters based on domain type. The ranks given by the evaluators and famousness scores are shown in Table 2.

Table 2. Person Ranks Given by Evaluators and Famousness Scores

	Prof. in K. Univ.					Prof. in DB Area				
	Prof. A	Prof. B	Prof. C	Prof. D	Prof. E	Prof. F	Prof. G	Prof. H	Prof. I	Prof. J
u_1	3	1	2	5	4	4	2	1	3	5
u_2	3	1	2	5	4	2	1	5	3	4
u_3	3	1	2	5	4	5	3	4	2	1
u_4	1	2	3	5	4	2	3	5	1	4
u_5	3	1	2	4	5	4	2	1	3	5
u_6	3	1	2	5	4	4	2	1	5	3
u_7	1	3	2	5	4	5	2	4	3	1
F_{111}	3	1	2	4	5	4	2	1	3	5
F_{100}	3	2	1	5	4	4	1	2	3	5
F_{010}	3	2	2	4	5	5	1	2	3	4
F_{001}	3	1	2	4	5	5	3	2	1	4

u_i stands for user i.
F_{111} stands for integrated famousness score whose weight parameters (α, β, γ) are (1,1,1).
F_{100} stands for integrated famousness score whose weight parameters (α, β, γ) are (1,0,0).
F_{010} stands for integrated famousness score whose weight parameters (α, β, γ) are (0,1,0).
F_{001} stands for integrated famousness score whose weight parameters (α, β, γ) are (0,0,1).

Table 3. Slide Ratios

		Person List 1 (Prof. in K. Univ.)					Person List 2 (Prof. in DB Area)				
		$s_r(1,5)$	$s_r(2,5)$	$s_r(3,5)$	$s_r(4,5)$	$S_R(3)$	$s_r(1,5)$	$s_r(2,5)$	$s_r(3,5)$	$s_r(4,5)$	$S_R(3)$
F_{111}	u_1	5/5	9/9	12/12	14/13	1	5/4	9/9	12/12	14/14	1.08
	u_2	5/5	9/9	12/12	14/13	1	5/5	9/7	12/10	14/11	1.16
	u_3	5/5	9/9	12/12	14/13	1	5/3	9/5	12/9	14/10	1.6
	u_4	5/4	9/7	12/12	14/14	1.18	5/3	9/4	12/9	14/13	1.75
	u_5	5/5	9/9	12/12	14/14	1	5/4	9/9	12/12	14/14	1.08
	u_6	5/5	9/9	12/12	14/13	1	5/4	9/9	12/10	14/12	1.15
	u_7	5/3	9/7	12/12	14/13	1.31	5/4	9/6	12/9	14/10	1.36
	average	-	-	-	-	1.07	-	-	-	-	1.31
F_{100}	u_1	5/4	9/9	12/12	14/14	1.08	5/4	9/9	12/12	14/14	1.08
	u_2	5/4	9/9	12/12	14/14	1.08	5/5	9/6	12/9	14/13	1.27
	u_3	5/4	9/9	12/12	14/14	1.08	5/3	9/5	12/9	14/14	1.6
	u_4	5/3	9/7	12/12	14/14	1.31	5/3	9/4	12/9	14/13	1.75
	u_5	5/4	9/9	12/12	14/13	1.08	5/4	9/9	12/12	14/13	1.08
	u_6	5/4	9/9	12/12	14/13	1.08	5/4	9/9	12/10	14/11	1.15
	u_7	5/4	9/7	12/12	14/14	1.18	5/4	9/6	12/9	14/10	1.36
	average	-	-	-	-	1.13	-	-	-	-	1.32
F_{010}	u_1	5/5	9/9	12/12	14/13	1	5/4	9/9	12/12	14/13	1.08
	u_2	5/5	9/9	12/12	14/13	1	5/5	9/6	12/9	14/11	1.28
	u_3	5/5	9/9	12/12	14/13	1	5/3	9/5	12/9	14/14	1.6
	u_4	5/4	9/7	12/12	14/14	1.18	5/3	9/4	12/9	14/11	1.75
	u_5	5/5	9/9	12/12	14/14	1	5/4	9/9	12/12	14/13	1.08
	u_6	5/5	9/9	12/12	14/13	1	5/4	9/9	12/10	14/13	1.15
	u_7	5/3	9/7	12/12	14/13	1.31	5/4	9/6	12/9	14/14	1.36
	average	-	-	-	-	1.07	-	-	-	-	1.32
F_{001}	u_1	5/5	9/9	12/12	14/13	1	5/3	9/7	12/12	14/14	1.31
	u_2	5/5	9/9	12/12	14/13	1	5/3	9/4	12/9	14/11	1.75
	u_3	5/5	9/9	12/12	14/13	1	5/4	9/6	12/9	14/14	1.36
	u_4	5/4	9/7	12/12	14/14	1.17	5/5	9/6	12/9	14/11	1.28
	u_5	5/5	9/9	12/12	14/14	1	5/3	9/8	12/12	14/13	1.26
	u_6	5/5	9/9	12/12	14/13	1	5/1	9/6	12/10	14/13	2.57
	u_7	5/3	9/7	12/12	14/13	1.32	5/3	9/4	12/8	14/13	1.81
	average	-	-	-	-	1.07	-	-	-	-	1.62

Table 4. Number of Matched Pages in Yahoo! Japan Web Search API Service

Prof. in K. Univ.					Prof. in DB Area				
Prof. A	Prof. B	Prof. C	Prof. D	Prof. E	Prof. F	Prof. G	Prof. H	Prof. I	Prof. J
294	1600	1630	1200	1480	294	1600	1580	7290	459
$S_R(3) = \frac{1.15+1.15+1.15+2.67+1.15+1.15+2.57+1.57}{7} \simeq 1.57$					$S_R(3) = \frac{1.32+1.38+1.29+1.15+1.32+2.67+1.43}{7} \simeq 1.51$				

To evaluate how quantitatively the ranks generated by our ranking method are close to those given by human evaluators, we use the slid ratio[8].

We assigned relevance scores to the professors depending on their ranks given by an evaluator. For example, if an evaluator ranked five persons in the order of

p_1, p_2, p_3, p_4 and p_5, we then give them relevance scores 5, 4, 3, 2 and 1, respectively. By using such that relevance scores, we computed the slide ratio as an evaluation measure of our ranking method. The slide ratio s_r is computed as follows.

$$s_r(m, n) = \frac{\sum_{j=n-m+1}^{j=n} j}{\sum_{i=1}^{i=m} r(i)} \qquad (27)$$

where $r(i)$ is the relevance score of the i-th person ranked by the famousness score. As we mentioned before, the relevance score depends on the ranks given by the evaluator. Here, n is the number of all given persons, and m is the ranking order given by the famousness score.

We computed the average slide ratio of the top k results as follows to evaluate our ranking method.

$$S_R(k) = \frac{1}{k} \cdot \sum_{i=1}^{i=k} s_r(i, n) \qquad (28)$$

where n is the number of given persons. We let $k = 3$ in our experimental evaluation.

The evaluation results related to the slide ratio are shown in Table 3. From the definition of S_R, it is obvious that $S_R(k) \geq 1$, and the smaller $S_R(k)$ is, the closer the ranks given by user and the famousness score will be.

As shown in Table 3, the average slide ratio of the professors in K. university is smaller than that of professors in the database area. One of the reasons for this difference is the evaluators are more familiar with their university's professors than those in database area who are at different universities. Because evaluators u_4 and u_7 are graduate students from other universities, they gave different ranks from the others. Moreover, the average slide ratios computed according to the ranks given by students (u_1, u_5 and u_6) who often go to conferences on databases area are smaller than the others. It is obvious that the ranks given by our ranking method are closer to those given by the evaluators who are familiar with the professors. That is, our people ranking method based on the notion of famousness is useful for ranking university professors.

Table 3 also shows that the average slide ratios of F_{111} are less or equal those of F_{100}, F_{010} and F_{001}. That is, the integrated score of famousness computed using all kinds of clusters may give better results. However, we need a further evaluation to verify this conclusion.

Table 4 shows the number of matched pages of these professors in the Yahoo! Japan Web search service. It is obvious that both the evaluators and our method did not rank the professors in the order of page numbers. If we simply rank these professors by using the number of matched pages, the average values of all evaluators' $S_R(3)$ could be computed to be 1.57 according to professors in K. university, and 1.51 according to professors in database area (see also Table 4). That is, although using the number of search results to rank people may be a simple and useful way sometimes, our method can generate ranks which are closer to those given by human evaluators.

5   Conclusion

In contrast to research on social network that analyzes the real world activities of people to utilize information on the Web, we focus on how to use the information available on the Web to support our daily activities.

In this paper, to support person search in the real world, we proposed a notion called famousness score to rank people according to a metadata analysis of search results. Intuitively, the famousness score of a person is computed from two aspects: 1) analysis within the searched Web pages related to that person and 2) comparison with other persons' search results. The basic idea is that persons having more Web pages from more areas and Web sites are more well-known than persons who have few pages from few Web sites and areas.

In contrast to conventional person search methods that map people to Web pages, our method is useful for ranking people and the experimental results validate this conclusion. It is notable that we can rank people with high accuracy by analyzing the metadata of search results returned by a search engine only.

Further studies on mechanisms for ranking people and evaluation are necessary. For instance, we plan to investigate the search results to improve the clustering methods. In the current work, we used the snippets to group Web pages. However, we could also use the full content of a Web page. To improve the computation of famousness score based on URL and type clusters, we plan to use more domain types to group Web pages into categories of private, TV, news-paper, and so on. The computation formula of the integrated famousness score is another future issue. A prototype system including visualization function will be developed in near future.

Acknowledgment

This research is partly supported by the research for the grant of Scientific Research (No.20700084, 20300042 and 20300036) made available by MEXT, Japan.

References

[1] Bekkerman, R., McCallum, A.: Disambiguating web appearances of people in a social network. In: Proc. of WWW 2005, pp. 463–470 (2005)

[2] Cheng, T., Chang, K.C.: Entity Search Engine: Towards Agile Best-Effort Information Integration over the Web. In: Proc. of CIDR 2007, pp. 108–113 (2007)

[3] Guha, R., Garg, A.: Disambiguating people in search. Stanford University (2004)

[4] Al-Kamha, R., Embley, D.W.: Grouping search-engine returned citations for person-name queries. In: Proc. of WIDM 2004, pp. 96–103 (2004)

[5] Kaseneci, G., Suchanek, F.M., Ifrim, G., Ramanath, M., Weikum, G.: NAGA: Searching and Ranking Knowledge. In: Proc. of ICDE 2008 (2008)

[6] Mann, G.S., Yarowsky, D.: Unsupervised personal name disambiguation. In: Proc. of CoNLL 2003 (2003)

[7] Ohshima, H., Nakamura, S., Tanaka, K.: SlothLib: A Programming Library for Research on Web Search (in Japanese). DBSJ Letters 6(1), 113–116 (2007)

[8] Pollack, S.M.: Measures for the comparison of information retrieval systems. American Documentation 19(4), 397–397 (1968)

[9] Suchanek, F., Kasneci, G., Weikum, G.: Yago: A Core of Semantic Knowledge. In: Proc. of WWW 2007 (2007)

[10] Wan, X., Gao, J., Li, M., Ding, B.: Person Resolution in Person Search Results: WebHawk. In: Proc. of CIKM 2005, pp. 163–170 (2005)

[11] Yahoo! Developer Network (2008), http://developer.yahoo.com/

[12] Yahoo! Japan Developer Network (2008), http://developer.yahoo.co.jp/search/

IWWUA 2008 Workshop PC Chairs' Message

Silvia Abrahão[1], Cristina Cachero[2], and Maristella Matera[3]

[1] Department of Information Systems and Computation,
Valencia University of Technology, Camino de Vera, s/n, 46022, Valencia, Spain
sabrahao@dsic.upv.es
[2] Department of Information Systems and Languages, University of Alicante
Campus de San Vicente del Raspeig. Apartado 99. 03080 Alicante, Spain
ccachero@dlsi.ua.es
[3] Dipartimento di Elettronica e Informazione, Politecnico di Milano
Via Ponzio, 34/5, 20133, Milano, Italy
matera@elet.polimi.it

Usability and accessibility are crucial factors in Web application development. The ease or difficulty that users experience with Web applications determines their success or failure. According to recent studies, an estimated 90% of Web sites and applications suffer from usability and/or accessibility problems. As user satisfaction has increased in importance, the need for usable and accessible Web applications has become more critical. To achieve usability for a Web product (e.g., a service, a model, a running application, a portal), the attributes of Web artefacts must be clearly defined. Otherwise, assessment of usability is left to the intuition or to the responsibility of people who are in charge of the process. In this sense, quality models (describing all the usability sub-characteristics, attributes and their relationships) should be built, and usability evaluation methods should be used during the requirements, design and implementation stages based on these models. Similarly, identifying the set of characteristics that make the Web more accessible for everybody, including those with disabilities, is necessary to systematize the way practitioners deal with accessibility issues.

The aforementioned motivations led us to organize the first edition of the International Workshop on Web Usability and Accessibility (IWWUA 2007) that was held in conjunction with the 8th International Conference on Web Information Systems Engineering (WISE), in Nancy, France, on December 2007. The success of this edition motivated us to organize a second edition of the workshop (IWWUA 2008) held in conjunction with the WISE 2008, in Auckland, New Zealand. The main purpose of the workshop is to discuss new trends in the evaluation of Web usability and accessibility, as well as to provide an international forum for information exchange on methodological, technical and theoretical aspects of the usability and accessibility of Web applications. These proceedings collect the papers presented at IWWUA 2008. All papers were peer-reviewed by three independent reviewers. The acceptance rate of the workshop was 50%. The majority of the papers aim at presenting novel approaches for Web usability and accessibility evaluation.

S. Hartmann et al. (Eds.): WISE 2008, LNCS 5176, pp. 61–62, 2008.
© Springer-Verlag Berlin Heidelberg 2008

In their paper, Insfran and Fernandez report on the results of a systematic review of usability evaluation in Web development. The goal was to investigate what usability evaluation methods have been employed in different research contexts to evaluate Web artefacts and how they have been employed. The paper by Koehnke et al. proposes the identification of semantic constructs in Web documents as a means to improve Web site accessibility. In this approach, semantic constructs represent building blocks of a Web document that correspond to a particular interaction activity of the Web site's usage scenario. Mckay and Burriss examine the usability problems resulting from the evaluation of novel Web software. In particular, they present an industrial case study with the purpose of improving an institutional repository from the end-user point-of-view.

Brajnik describes a method called MAMBO (MAnually Measuring Barries Of accessibility) for measuring barriers of accessibility. An experimentation with the method is also described based on the analysis of 14 accessibility reports. Molina, Pardillo and Toval discuss the crucial issue of eliciting usability and accessibility requirements. To address this issue, they propose a Web engineering requirements meta-model that can be smoothly integrated with existing Web engineering proposals in order to reinforce the first steps of Web development. Finally, the paper by Xiong et al. presents an ontology-based approach for dealing with usability and accessibility guidelines for Web applications.

We would like to thank all the authors for submitting their papers to the Workshop and contributing to shape up such a rich program, the members of the Workshop Program Committee for their efforts in the reviewing process, and the WISE organizers for their support and assistance in the production of the proceedings. We are also grateful to Giorgio Brajnik from University of Udine, Italy, who agreed to give a keynote speech on *Beyond Conformance: The role of Accessibility Evaluation Methods*. Finally, we would like to thank the CALIPSO (http://alarcos.inf-cr.uclm.es/calipso) and the MAUSE (Towards the Maturation of IT Usability Evaluation www.cost294.org) projects for sponsoring the workshop.

Beyond Conformance:
The Role of Accessibility Evaluation Methods

Giorgio Brajnik

Dip. di Matematica e Informatica
Università di Udine
Italy
www.dimi.uniud.it/giorgio

1 Introduction

The topic I want to address is the role that accessibility evaluation methods can play in helping the transition from accessibility viewed as standard conformance, to a user-centered accessibility. As we will see, this change sets additional requirements on how evaluations of websites should be carried out.

This paper first discusses different problems that occur while dealing with accessibility. We will see that different people have radically different views of accessibility and how it should be assessed.

The first requirement is a clear definition of what accessibility is and how it should be assessed. The accessibility model discussed in Section 2.1 has precisely this role.

Several existing evaluation methods are then reviewed and discussed, a simple taxonomy is presented, and differences that occur when evaluating accessibility rather than usability are pinpointed.

1.1 Problems in Managing Web Accessibility

As discussed by Kelly et al. [22], the W3C/WAI model of accessibility aims at *universal accessibility*, it assumes that website conformance to WCAG (Web Content Accessibility Guidelines) is the key precondition to that, and it hypothesizes that accessibility is entailed by a conformant website if two other conditions are met. Namely, that the tools used by the web developer (including CMSs) are conformant to ATAG (Authoring Tools Accessibility Guidelines), and that browser and assistive technology used by the end user are conformant to UAAG (User Agent Accessibility Guidelines). However, since both these two conditions are not under control of the web developer, the conclusion is that the developer cannot guarantee accessibility, whatever efforts s/he may put it.

In fact, empirical evidence shows that the link between conformance and accessibility is missing, *i.e.* even conformant websites may fail in being accessible [13, 29].

Confusion exists regarding the methods to use. For example, the current Italian regulation for web accessibility [19] specifies a number of technical requirements similar to WCAG 1.0 and Section 508 points. However, in order to certify

S. Hartmann et al. (Eds.): WISE 2008, LNCS 5176, pp. 63–80, 2008.

accessibility evaluators have to perform a *cognitive walkthrough*, that is an analytical method generally used for early-on usability investigations, whose effectiveness as a method for accessibility evaluations is yet unproven. In addition, the regulation specifies 12 general usability principles that are generally employed with *heuristic evaluation*. It also requires that evaluators classify identified problems into 5 severity levels, without specifying how severity should be determined. It then suggests using an empirical method that again has no proved effectiveness (*i.e. subjective assessments*) and finally it requires that evaluators compute a final score for the site on the basis of mean averages of severity levels (an ineffective aggregation technique of ordinal variables). Although such a regulation sets a certification framework for web accessibility, in my view it is unlikely to succeed because of extreme subjectivity and variability, poor practicality and measure-theoretical shortcomings.

As evidence of further confusion, consider the Target legal case in the U.S.A.[1]. The National Federation for the Blind (NFB) claimed that `target.com` is not accessible since some NFB's witnesses gave up when using the site; on the other hand, Target's witnesses testified that they were able to navigate, shop and that they actually enjoyed it; in addition, an NFB's expert declared in court that `target.com` fails to address accessibility since:

> ... 15 of the site's pages were analyzed: six top-level pages as well as nine pages that had to be navigated in order to complete a purchase. In those fifteen pages, alt-text was missing on 219 active images (links); none of the form controls were properly labeled; and there was no accommodation for screen reader or keyboard navigation, such as skip links or HTML headings.

Finally, the Court concluded that the question of the accessibility of `target.com` was not decided and so it refused to grant a preliminary injunction.

We can see that there is substantial variability, and lack of standardization, in the way pages were selected, in the way accessibility was investigated, and in the way a conclusion was drawn. Witnesses of one side were referring to user performance indicators, the others to conformance features.

Additional evidence exists showing that accessibility evaluation based on a sample of pages (sampling is necessary for all but trivial websites) can be affected by the criteria used to select the sample. There is interdependence between the sampling criteria and the purpose of the accessibility analysis [8], leading to large differences in accuracy. If the evaluation aims at conformance, then the most frequently used sampling criterion (selecting predefined pages: home, contact, site map, etc.) may lead up to a 38% inaccuracy rate, *i.e.* 38% of the checkpoints may be wrongly estimated.

My claim is that to change this state of things we have to focus on how to standardize methods, and through them aim at an accessibility that is sustainable; in other words, we need to shape and establish effective accessibility processes that can be sustained mainly by their own return on investment.

---

[1] See `www.jimthatcher.com/law-target.htm` for details.

At least two issues have to be addressed. First, accessibility evaluations have to produce sets of accessibility problems that are prioritized by their impact: in other words, evaluations should identify problems whose solution makes a difference in accessibility as viewed by stakeholders. Therefore, evaluators and developers can focus on these problems first, and optimize their resources. Secondly, accessibility processes (taking place when conceiving, developing, maintaining, revamping websites) should be effective and efficient, and these properties should be the result of scientific investigations. When these two conditions are met, then accessibility methods can be compared and chosen on an informed basis, and this will lead to more accessible websites/web applications that in turn will positively affect key performance indicators related to the underlying business the website should support.

The relation between accessibility and usability is also controversial. According to Thatcher et al. [36], accessibility problems affect only disabled people and have no effect on others. Petrie and Kheir [29] mention that they noticed that disabled and non-disabled people often encounter the same problems, but are affected by them differently. Slatin and Lewis [33] performed an experiment aimed at determining whether accessibility features of a website positively affect non-disabled users. While vision-impaired subjects improved their success rate and productivity when using the accessible version of a website, no statistically significant difference was found for non-disabled subjects. The implication of this study is that accessibility does not necessarily lead to higher usability. On the basis of a comparative experiment between vision-impaired and sighted users, Petrie and Kheir [29] reach the conclusion that the problems faced by the two groups were overlapping, but the overlap was small, and the majority of the problems were found only by disabled users. This study did not detect significant differences in the severity of the problems found by the two groups.

We can see that accessibility and usability are different; my view is that currently a good accessibility model is missing.

2   The Role of Accessibility Models

2.1   An Accessibility Model

A model of accessibility specifies what accessibility is, how it is achieved and managed, and which boundary conditions can influence it. A model not only helps to plan and perform activities like diagnosing the defects of a website, monitoring it and comparing it to other websites, measuring its accessibility level to determine whether it is conformant to certain standards.

More specifically the accessibility model I propose addresses the following questions and comprises the following components.

Properties. Which properties should be central in the notion of accessibility? To respond to this question, if we look at definitions that were proposed for accessibility in the past (see Table 1), it's clear that very different properties are taken into account. Some definitions focus on user performance

Table 1. Existing definitions of accessibility

Source	A website is accessible if ...
W3C/WAI [45]	... its pages transform gracefully despite constraints caused by physical, sensory, and cognitive disabilities, work constraints, and technological barriers, and its content is understandable and navigable.
Slatin and Rush [34], U.S. Dept. of Justice [39]	... disabled people can use it with the same effectiveness as non-disabled people.
Thatcher et al. [35]	... it is effective, efficient and satisfactory for more people in more situations.
ISO [18]	... it is usable by people with the widest range of capabilities
Italian Parliament [20]	... deploys services and information so that they can be exploited with no discrimination also by disabled persons.
Thatcher et al. [36]	... people with disabilities can perceive it, understand it, navigate it and interact with it.
Petrie and Kheir [29]	... it can be used by specific users with specific disabilities to achieve specific goals with effectiveness, efficiency and satisfaction in a specific context of use.
W3C/WAI [46]	... its content must be perceivable to each user; user interface components in its content must be operable by each user; content and controls must be understandable to each user; content must be robust enough to work with current and future technologies.
College of Design, North Carolina University [10]	... it is usable by all people, to the greatest extent possible, without the need for adaptation or specialized design.

indicators that can be experimentally measured (*e.g.* effectiveness, usability^2), one definition sets appropriate relative levels (*e.g.* same effectiveness), other definitions focus on properties that are more difficult to define and measure (*e.g.* navigability, understandability, exploitation); sometimes even properties unrelated to user-performance properties are considered (*e.g.* robustness, degradation). The last definition refers to Universal Design, which is often considered to be the same as accessibility. Such a definition excludes many contextual elements that are central in the definition of usability, reducing in such a way the power of AEMs, as we will see below.

In this paper I will assume that a website is accessible when

specific users with specific disabilities can use it to achieve specific goals with the same effectiveness, safety and security as non-disabled people.

---

^{2} *Effectiveness* is the accuracy and completeness levels that can achieved by specified users when aiming at specified goals under specified conditions. *Usability* is the effectiveness, productivity, satisfaction and security with which specified users can achieve specified goals under specified conditions; *productivity* is related to the resources expended (time, effort, skills, infrastructure) in achieving those goals at given levels of effectiveness (ISO 9241). See books like [7, 31] for relevant metrics.

This definition points to measurable user-performance parameters, sets viable, relative thresholds and restricts the claim to certain users and goals.

Context. Which additional factors influence accessibility and how can they be detected, isolated and controlled?

As we move from a viewpoint were accessibility is equated to conformance to some standard, to a view were accessibility becomes user-centered, then context plays an increasingly important role and needs to be considered whenever accessibility is evaluated. Context should provide enough information to enable evaluators to determine possible hurdles for users and their consequences.

Ideally context should address the "who", "what" and "how" questions: (i) the type of user disability, (ii) the experience level in using the browser, the Web, the assistive technology, and possibly the specific website and domain of operation, (iii) the short-term user goals, (iv) the physical environment the user is working in (posture, light and noise conditions), (v) input and output devices and interaction modalities (media used, possible user actions and operations, user agents, assistive technologies and infrastructure).

Methods. Given that we know on what properties to focus, and how to characterize boundary conditions, how are we going to detect and measure these properties accurately and reliably?

This ingredient of the accessibility model comprises techniques, methods and methodologies used to evaluate, assess and manage accessibility. As we will see later on, there are a number of evaluation methods usually put in operation for accessibility; some of them are adapted from usability methods, others are specific to accessibility. Nonetheless, few studies are available to shed light on how well these methods work for accessibility, making the choice of the evaluation method and the choice of the metric to use for measuring accessibility very uncertain.

2.2   The Importance of Context

Context is more crucial for accessibility than it is for usability. Besides being dependent on users' experience, goals and physical environment, accessibility of a website depends also on the platform that's being used. It is the engine of a transformation process that is not under control of the web developer. In fact, accessibility of digital media requires a number of transformations to occur automatically, concerning the *expression* of the *content* of the website [1, 7]. Content is the meaning that a person (*e.g.* visitor, developer, evaluator) associates to perceivable elements of a web page, which constitute its expression. See Table 2 for a brief taxonomy. Examples of transformations include text that could be read aloud; images that could be "transformed" into spoken words (via their textual equivalents); scenes of a video that could be enriched with textual captions describing them; audio content that could be transformed into textual transcripts; changes in font attributes.

These changes in expression involve *inter-media* transformations (*e.g.* text to spoken words), *intra-media* transformations (*e.g.* by changing the geometric

Table 2. Taxonomy of content and expression

Content elements	
Interest information	Concepts, questions, answers that can satisfy users goals
Bearing information	Location information (*e.g.* breadcrumbs, headings), direction information (*e.g.* link labels)
Access information	Supports user actions (*e.g.* navigation bars, sequential paging, filters)
Functional information	Provided by users and necessary to achieve the goal (*e.g.* address data provided when completing an on-line order)
Expression elements	
Expression media	Text, image, sound, video
Expression style	Font, size, colors, texture, orientation, ...
Compositional structure	Spatial, temporal, spatio-temporal, or hyper-medial

properties of space when using a screen magnifier to enlarge the screen, or when changing the text size), *temporal* transformations through new synchronization of events (*e.g.* by using audio signals to notify a user of a screen reader that a certain feedback message has appeared in a location that differs from the current focus of interaction) or slow-down of an animation/simulation; finally *de-contextualization* of information occurs (*e.g.* when the user of a screen reader extracts and lists all the links in a page, so that each link is rendered out of its original context).

Some of these transformations affect interaction modality. For example, new operations are made available, like the ability to extract and scan links in a page, or page headings, or the ability to jump directly to the content of the page, or the ability to move back and forth through items of a list. In a sense, this perspective on transformations occurring for the sake of accessibility is close to the notions of *plasticity* and *retargeting* of the user interface[3]. Notice however that these transformations occur on the fly and solely on users' platform.

As a consequence of the definition of accessibility I gave before, a website is accessible only if the transformation of web pages from one expression to another is such that *invariance of content* is preserved, in specified contexts. In other words, regardless of the expression and interaction modality used by the visitor and within given contexts, the same content is rendered, reaching the same level of effectiveness. Finally, many diverse transformations have to be enabled for each "target" interface required by the assistive technology that is considered in the model.

Invariance of content holds if a number of enabling conditions are met. First, the platform should support all required transformations and the technologies

---

[3] Plasticity is the ability of the system to produce a user interface that is adapted to the device being used and possible context of use. Retargeting means to statically analyze a web page, to automatically derive an abstract user interface (*e.g.* by inferring the existence of an abstract object called "RadioButton"), to transform such an abstract interface into the abstract interface for another platform (*e.g.* on a mobile device the "RadioButton" object is mapped into a "Listbox"), and finally to generate an appropriate and running user interface on the selected platform [9, 38].

used by the website (*e.g.* HTML, CSS, JavaScript, SMIL, Flash, PDF). Second, website developers need to provide the required redundant expression in the different media that might be needed (*e.g.* textual descriptions of video scenes). They also have to provide specifications to support transformation of expression (*e.g.* synchronization constraints so that captions are rendered at the right time). Thus, from the perspective of authors, accessibility requires them to clearly identify all the content units and make sure that (i) interest and bearing information can be transformed into all possible media that might be available in users platforms, (ii) that the transformations are complete (*e.g.*, all bearing information is transformed) and (iii) that operability is guaranteed, *i.e.* all functional information and access information can be operated in the transformed interaction modality.

Context can affect all three of these conditions, which don't usually occur when dealing with usability; this is why it plays a more important role in accessibility and why it poses more challenges to evaluators and developers. Therefore, in order to be accurate and produce relatively standardized results, evaluation methods need to consider context.

3   Accessibility Evaluation Methods

With *accessibility evaluation method* (AEM) I mean a procedure aimed at finding accessibility problems, such as guideline violations, failure modes, defects[4], or user performance indexes. More specifically an AEM:

1. prescribes which steps, which decisions, which criteria should be used under which conditions, so that accessibility problems can be detected;
2. may prescribe how to classify and rate problems (in terms of severity, priority, or else);
3. may prescribe how to aggregate data about problems, as well as how to describe and report them;
4. may prescribe how to select web pages for evaluation.

3.1   A Taxonomy of Accessibility Evaluation Methods

Several methods can be used to find accessibility problems; they are reviewed in Section 3.2. Before discussing each of them in detail, however, I provide a taxonomy highlighting criteria that can be useful to contrast them; see Figure 1. Some of the criteria illustrated below were discussed also by Hartson et al. [15].

---

[4] *Failure mode* to the way in which the interaction fails; the *defect* is the reason for the failure, its cause; *effects* are the negative consequences of the failure mode. In this context, an *error* is a wrong design/implementation decision taken by developers. For example, a failure mode may be the inability of a screen reader user to swiftly navigate around elements of a web page; a corresponding defect may be the absence of *skip links* links and of page headings; effects include a reduction of user productivity, satisfaction and a dramatic reduction of effectiveness if the user — each time a new page is reloaded — has to repeatedly press the TAB key to get to the desired content of the page.

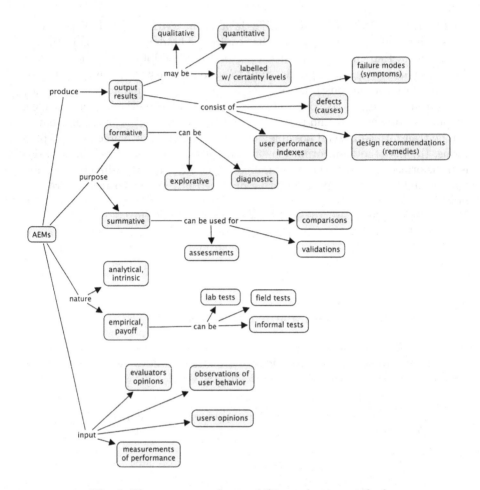

Fig. 1. The taxonomy of accessibility evaluation methods

Methods can be *analytic, empirical,* or both: the former are based on inspections of web pages usually carried out by experienced evaluators, without putting pages in a real work context. Empirical methods, sometimes used to perform so called *payoff* evaluations, require that an interaction takes place between users and the studied website. Empirical methods may be *laboratory based,* when potential disturbances to the user interaction are reduced to the minimum, or *informal ones,* when the strict "aseptic" conditions are not needed.

Methods differ also according to the *information used* to derive accessibility problems: some methods are based on observations of the behavior of users; others on opinions expressed by users or by evaluators.

In terms of results produced, AEMs can yield descriptions of *failure modes,* or may produce also corresponding *defects* and even design recommendations, *i.e. solutions.* Some methods produce synthetic measures of users' performance indicators, called payoff functions (*e.g.* effectiveness measured as success rate).

Results can be qualitative or quantitative; they may also be qualified with a confidence degree (like the probability of being a wrong result) and can support generalization to a wider population of Web users and/or to a wider set of conditions.

Regarding their purpose, methods can be used to perform *formative* evaluations aimed at identifying lists of problems. These methods can be used to explore failure modes that are the accessibility obstacles to a smooth interaction (*explorative* evaluations); they can also identify defects and solutions so that problems can be fixed (*diagnostic* evaluations). Formative evaluations are (or should be) used during the development, supporting iterative design. *Summative* evaluations, on the other hand, are carried out to assess the accessibility level of an interface; differently than formative evaluations, summative AEMs may produce only aggregated results regarding user performance measures (*e.g.* global effectiveness, productivity, user satisfaction figures). They can be deployed to estimate the accessibility of an interface (*assessment*), to *validate* it or to *compare* one interface against different versions or different systems. If summative evaluations are used for comparing different websites that may be used within different contexts, which is what happens when conformance is assessed, then some sort of standards of conditions should be defined. Only in these cases the results of the evaluation can be compared safely.

3.2   Review of Existing Methods

Although close to usability, accessibility has its own evaluation methods, and few are in common. I will briefly review the most typical ones, highlighting their benefits and disadvantages. Table 3 summarizes this discussion.

Ideally, a good method is a dependable tool that yields accurate predictions of all the accessibility problems that may occur in a website. This is why methods are compared in terms of such criteria as *correctness* (the percentage of reported problems that are true problems), *sensitivity* (the percentage of the true problems being reported), *reliability* (the extent to which independent evaluations produce the same results), *efficiency* (the amount of resources expended to carry out an evaluation that leads to specified levels of effectiveness and usefulness), *usefulness* (the effectiveness and usability of the produced results) and the method's *usability* (how easily it can be understood, learned and remembered by evaluators); for more details the reader is referred to [14, 15, 17, 23, 32].

Conformance Reviews. Called also *expert, standards, or guidelines review* or *manual inspection* [16, 42], this is by far the most widely used AEM [12]. It is based on checking whether a page satisfies a checklist of criteria. It is an analytic method, based on evaluators' opinions, producing failure modes (in the form of violated checkpoints) possibly with defects and solutions. Conformance reviews are used in both formative and summative evaluations: the former when defects are diagnosed in order to be fixed, the latter when a formal conformance statement is needed (for assessment, validation or comparison).

The method often entails the following steps [19, 42]: (i) determining an appropriate sample of web pages (including different sorts of tables, forms, images and scripts), (ii) running markup validators on selected pages, (iii) cross-checking selected pages against all applicable checkpoints, (iv) examining selected pages with a range of graphical, textual and voice browsers, and finally (v) summarizing the results.

Benefits of this method include the ability to identify a large range of diverse problems for a diverse audience (albeit this depends on the quality of the underlying checkpoints); it is relatively cost-effective, especially when coupled with automatic testing tools, and, by being diagnostic in nature, it can be used to identify the defects underlying the checkpoint violations, hence assisting those who have to fix them.

Conformance reviews are dependent on the chosen checklist, that range from standards issued by international bodies (like the Web Content Accessibility Guidelines, WCAG, published by the W3C), to national or state-level guidelines, like [19, 40], to individual organizations guidelines (like those issued by IBM, SUN or SAP, for example). Guidelines of course affect the quality of this method. As discussed in [21], WCAG 1.0 guidelines suffer from their theoretical nature, dependency on other guidelines, ambiguity, complexity, their closed nature and by some logical flaws.

The study [13] found fundamental limits of conformance review with respect to WCAG 1.0: "as many as 45% of the problems experienced by the user group were not a violation of any checkpoint, and would not have been detected without user testing". The study identified gaps in the guidelines such as reducing deep structures in websites, improving search mechanisms, preserving links to home pages, reducing the number of existing links; the study suggested also a reordering of checkpoint priorities.

The project by Theofanos and Redish [37], performed as a field study with 16 users over a period of months, highlights several guidelines that address usability for screen reader users and that go beyond conformance: starting links with significant words, rewording questions with the main topic first, writing "home page" rather than "homepage", avoiding page refresh, synchronizing alt-text with text in the page, are some of the 32 suggested (additional) guidelines. This holds true also for the guidelines proposed by Nielsen Norman Group [27, 28]. In agreement with [24], the study found that an additional shortcoming of conformance review is the large number of possible guidelines and principles to choose from.

Rating the severity of problems through analytical methods appears also to be a source of methodological weakness. Petrie and Kheir [29] showed that while participants and experimenters agreed substantially on assigning severities to problems found via empirical methods, the agreement on these severities with checkpoint priorities in WCAG 1.0 was extremely poor. The same happened with respect to usability guidelines. This result suggests that it's extremely inaccurate to use fixed predefined priorities/severities. For example, few of the images in a website that lack an appropriate alternative text are a true barrier: most of the images are used for emotional purposes, which in textual alternatives would be lost anyway. But an important function of an evaluator is to find out what the

Table 3. Summary of pros and cons of AEMs

pros	cons
conformance review (CR)	
− low cost − diagnostic − suitable for formative and summative eval. − identifies a large spectrum of problems	− requires skilled evaluators − does not support the evaluator in assigning severities − not practical with lots of pages − conformance does not mean accessibility − unable to catch important usability problems − guidelines may be complex to read, too abstract, too many − with inexperienced evaluators, it is less effective than other methods
subjective assessment (SA)	
− low cost, low difficulty − can be done remotely − good correctness	− not systematic (problems and/or pages) − highly dependent on users' experience − users may not be aware of certain problems − poor description of problems − low thoroughness − requires users with different disabilities
screening techniques (ST)	
− low cost − suitable for formative eval.	− time consuming for web developers − singles out certain disabilities − yields developers opinions − highly dependent on developers experience − cannot be used for summative eval.
barrier walkthrough (BW)	
− low cost, low difficulty − supports learning − higher correctness than CR − yields severity ratings	− lower sensitivity than CR − dependent on evaluators experience − less reliable than CR
user testing (UT)	
− highlights important problems − leads to correct severity ratings	− higher cost than analytical methods − logistics is complicated − mixes accessibility with usability problems − should not be done remotely

consequences of such defects are: this, however, can be done only if appropriate use scenarios are considered. Conformance review does not prescribe how to choose scenarios nor how to rate the defect, except for static priorities that cannot reflect specific usage scenarios.

Automated Tests. Though closely related to conformance reviews and very popular, methods that are based exclusively on automated testing tools, like those listed in [44], should not be considered *evaluation* methods. The reason is that these tools have to rely on heuristics to determine violations of several checkpoints. The quality of these heuristics is not satisfying: in a previous study [4] we found false positives to be up to 33% and false negatives to 35%; Thatcher et al. [36] found that of 40 different benchmark tests the best and worst of six tools passed respectively 23 and 38 tests, a failure rate between 5% and 42%.

Therefore, using *only* automated tools is not by itself a viable solution to the problem of evaluating accessibility. The W3C/WAI puts it nicely[5]: "Web accessibility evaluation tools can not determine the accessibility of Web sites, they can only assist in doing so." On the other hand, because they are systematic and fast, tools yield other important advantages, namely effectiveness, productivity, and wide coverage of web pages. They are therefore an important option in the portfolio of a careful evaluator.

An interesting approach is to consider context by tailoring the automatic evaluation to peculiarities of certain types of users in the form of personal accessibility profiles; see for example [41].

Screening Techniques. These are informal empirical techniques based on using an interface in a way that some sensory, motor or cognitive capabilities are artificially reduced [16]. For example, an evaluator would use a website through a screen reader with the monitor turned off; or after unplugging the mouse; or by using the mouse with the left hand (for a right-handed person). After carefully selecting the screening conditions so that they match the characteristics of the target population, the evaluator explores the website and tries to accomplish selected goals. Hindrances that occur during such a process are accessibility problems that these empirical, informal, explorative techniques can detect.

Their benefits include the relatively low performing costs (the evaluator has to install and learn how to use a number of assistive technologies, but this is a one-time cost). However it is a method that is not systematic, and we should expect it to show low effectiveness since it depends heavily on the experience level of the evaluator in using the assistive technology, which rarely would match the experience of users.

Mankoff et al. [24] report that web developers using the screen monitor together with the screen reader were able to reach good levels of sensitivity which are comparable to conformance review.

Subjective Assessments. Rubin [31] calls them *self-reporting methodologies*. When applying this method, the evaluator involves a panel of users (sharing

---

[5] www.w3.org/WAI/eval/selectingtools.html

characteristics with the reference audience), instructs them to explore and use a given website, which they do by individually or jointly with other users. Then the users are interviewed, directly or through a questionnaire — that can be submitted during the usage —, providing feedback on what worked for them and what did not. The evaluator extracts the list of accessibility problems from this body of self-reported user opinions. Depending on users' experience in accessibility, the method can be not only empirical, but also analytical and diagnostic; it is based on users' opinions, and can yield failure modes, defects and possible solutions. Therefore it can be adopted for explorative or diagnostic formative evaluations.

Its benefits include the low cost, the fact that it does not require experienced evaluators, and the ability to carry it out remotely in space and time (*i.e.* asynchronously). In addition, participants may be allowed to explore areas of the website that most suit them, with corresponding increase of their motivation in using the website.

However there are important drawbacks: the method is very unsystematic, not only regarding the pages that are being tested, but also the criteria used to evaluate them. In addition, different users with different experience levels and different attitudes will report very different things about the same page. Subjects have to remember what happened during an interaction, they often rationalize their behavior, and may be distracted without being aware of it. Mankoff et al. [24] discovered that this method ranks well in terms of correctness, but poorly in terms of sensitivity when compared to conformance review and to the screening technique mentioned above.

Barrier Walkthrough. The barrier walkthrough method is an analytical technique based on heuristic walkthrough [32] that I proposed in [3, 5]. An evaluator has to consider a number of predefined possible barriers which are interpretations and extensions of well known accessibility principles; they are assessed in a context which potentially includes the elements described in Section 2.1 so that appropriate conclusions about user effectiveness, productivity, satisfaction, and safety can be drawn, and severity scores can be derived. For BW, context comprises user categories (like blind persons), website usage scenarios (like using a given screen reader), and user goals (corresponding to *use cases*, like submitting an IRS form). An *accessibility barrier* is any condition that makes it difficult for people to achieve a goal when using the website in the specified context. A barrier is a failure mode of the website, described in terms of (i) the user category involved, (ii) the type of assistive technology being used, (iii) the goal that is being hindered, (iv) the features of the pages that raise the barrier, and (v) further effects of the barrier on payoff functions.

The BW method prescribes that severity is graded on a 1–2–3 scale (minor, major, critical), and is a function of *impact* (the degree to which the user goal cannot be achieved within the considered context) and *persistence* (the number of times the barrier shows up while a user is trying to achieve that goal). Therefore the same type of barrier may be rated with different severities in different contexts; for example, a missing *skip-links* link may turn out to be a nuisance for a blind user reading a page that has little preliminary stuff, while the same defect may show a higher severity within a page that does a server refresh whenever the user interacts with a sequence of select boxes. Compared to suggestions on how to rate problems given by Nielsen [26], I believe that "cosmetic" problems should not be considered when evaluating accessibility.

Potential barriers to be considered are derived by interpretation of relevant guidelines and principles [13, 37, 43]. A complete list can be found in [5].

We should expect two major benefits of BW compared to conformance review: by listing possible barriers grouped by disability type, evaluators should be more constrained in determining whether the barrier actually occurs. Secondly, by forcing evaluators to consider usage scenarios, an appropriate context is available to rate severity of the problems found.

In fact, a preliminary experimental evaluation of the BW method [3] showed that this method is more effective than conformance reviews in finding more severe problems and in reducing false positives; however, it is less effective in finding all the possible accessibility problems. Some of these results agree with findings reported by Sears [32], who compared heuristic walkthrough with other inspection-based usability evaluation methods, heuristic evaluation and cognitive walkthrough.

Other studies showed how BW can be used as a basis for measuring the accessibility level of a website rather than measuring the conformance level. In particular [6] illustrates how the output of an accessibility testing tool can be sampled so that an assessment similar to BW is performed by one or more judges. On the basis of these sampled barriers, estimates of tool errors and of the accessibility of the website can be computed. These computations can be performed also on conformance review reports, again on the basis of a judging step based on BW [2].

User Testing. Even though empirical methods like laboratory and field testing can in principle be used for evaluating accessibility, the method more often chosen is the lightweight *informal user testing* through the think-aloud protocol [11, 16, 25, 26, 31]. Once a panel of users is selected (representing the target audience in terms of disability, user roles with respect to the website, experience levels in the Internet, in assistive technologies, and in the specific domain and website), they are required to perform given tasks while being observed and being asked to think aloud. In the end, from notes, audio and video recording taken during the test run, evaluators generate the list of problems and assign severity levels.

To ensure effectiveness, the protocol used by evaluators to identify problems should be carefully defined to reduce what Hertzum and Jacobsen [17] call the "evaluator effect", which influences the kind of problems that are detected, at which level of abstractions, and how they are rated for severity. Furthermore, users should be asked to use applications and assistive technologies they are familiar with, and they should be screened according to their level of experience in using these tools.

One benefit of user testing is important [23]: its capability to accurately identify usability problems that are usually experienced by real users, and that have potentially catastrophic consequences. Conversely, this method is not suitable to identify low-severity problems.

More important drawbacks include its higher costs compared to analytic methods and its inability to highlight defects in addition to failure modes. Furthermore, problems may be missed if predefined scenarios are not well chosen or if user disabilities, experience levels or roles are not representative of the target audience. In addition, given users' requirement in terms of appropriate assistive technology and room facilities, setting up a user testing session with disabled participants may be challenging; similarly for recruiting a group that represents the target audience. Results of performing user testing are likely to be a set of usability problems that are general for all the users of the website, rather than being specific for disabled persons (e.g. a misleading link label). In other words, the method is likely to identify a number of true problems, but irrelevant with respect to accessibility (as defined in Section 2.1).

Finally, as reported by Petrie et al. [30], remote user testing for accessibility eases the logistic difficulties, but raises additional issues concerning the validity of results. In two studies comparing local v.s. remote user testing, they found out that asynchronous remote user testing, where users work at home on given tasks and websites, and take notes of problems, is a method that can be used with some care for summative evaluations, but is unlikely to be useful for formative evaluations. The reason is that the level of details used to describe problems is much higher when the evaluator observes, and perhaps, challenges the user. Secondarily, care must be payed so that reliable data is gathered concerning the success levels achieved by users. The problem is that often users are not aware of missing part of the goal.

As a final remark, note that all usability evaluation methods can be used to assess accessibility, provided it is understood that in such cases accessibility really means "usability with respect to disabled users and the particular operating conditions determined by the platform used". When this is true then heuristic evaluations and walkthroughs, cognitive and pluralistic walkthroughs, user tests of different sorts can all be used. For analytic methods, the list of principles, guidelines, tasks and basic questions is exactly the same as when dealing with people with no disabilities (e.g. the guidelines proposed in [26]).

4   Conclusions

We have seen an accessibility model that clearly defines what accessibility is, how to assess it, and how to represent context. To distinguish accessibility from usability, accessibility should aim at non discriminating users in terms of what they can achieve; accessibility should focus on websites capable of providing equal levels of effectiveness, safety and security in specified contexts. Context should include descriptions of "who" is going to use the website (type of disability, experience level in the Internet, in the specific assistive technology, in using the specific website and its domain), for doing "what" (user goals), and "how" (physical environment and interaction modalities).

Context is necessary when moving from conformance to accessibility, and it has to be considered also in evaluation methods. The methods I reviewed treat context differently. It is virtually absent or very general when performing conformance reviews; it is usually implicitly defined in subjective assessments and screening techniques; it is explicitly characterized in barrier walkthroughs and in user testing. This, in my view, reflects how applicable a method is for evaluating accessibility and affects its correctness, sensitivity and reliability.

More work is needed to provide additional evidence of advantages and disadvantages of methods; but I believe that the adoption of the model and more focus on context would help the web accessibility community to resolve the kind of problems affecting accessibility.

References

[1] Brajnik, G.: Modeling content and expression of learning objects in multimodal learning management systems. In: HCI International 2007, FUITEL: Future Interfaces in Technology Enhanced Learning, Beijing, China (July 2007)

[2] Brajnik, G.: Measuring web accessibility by estimating severity of barriers. In: 2nd International Workshop on Web Usability and Accessibility IWWUA 2008, Auckland, New Zealand (September 2008)

[3] Brajnik, G.: Web Accessibility Testing: When the Method is the Culprit. In: Miesenberger, K., Klaus, J., Zagler, W., Karshmer, A.I. (eds.) ICCHP 2006. LNCS, vol. 4061, pp. 156–163. Springer, Heidelberg (2006)

[4] Brajnik, G.: Comparing accessibility evaluation tools: a method for tool effectiveness. Int. Journal on Universal Access in the Information Society 3(3-4), 252–263 (2004)

[5] Brajnik, G.: Web accessibility testing with barriers walkthrough (March 2006) (Visited May 2008), http://www.dimi.uniud.it/giorgio/projects/bw

[6] Brajnik, G., Lomuscio, R.: SAMBA: a semi-automatic method for measuring barriers of accessibility. In: Trewin, S., Pontelli, E. (eds.) 9th Int. ACM SIGACCESS Conference on Computers and Accessibility, ASSETS, Tempe, AZ. ACM Press, New York (2007)

[7] Brajnik, G., Toppano, E.: Creare siti web multimediali: fondamenti di analisi e progettazione, Italy. Addison-Wesley/Pearson Education (2007)

[8] Brajnik, G., Mulas, A., Pitton, C.: Effects of sampling methods on web accessibility evaluations. In: Trewin, S., Pontelli, E. (eds.) 9th Int. ACM SIGACCESS Conference on Computers and Accessibility, ASSETS, Tempe, AZ. ACM Press, New York (2007)

[9] Buillon, L., Vanderdonckt, J.: Retargeting web pages on other computing platforms with vaquita. In: van Deursen, Burd, A. (eds.) Proc. of IEEE Working Conf. on Reverse Engineering WCRE 2002, Richmond, October 2002, pp. 339–348. IEEE Computer Society Press, Los Alamitos (2002)

[10] College of Design, North Carolina University. Principles of Universal Design. The Center for Universal Design (Febuary 1997) (Visited May 2008), http://www.design.ncsu.edu/cud/about_ud/udprincipleshtmlformat.html

[11] Coyne, K.P., Nielsen, J.: How to conduct usability evaluations for accessibility: methodology guidelines for testing websites and intranets with users who use assistive technology. Nielsen Norman Group (October 2001), http://www.nngroup.com/reports/accessibility/testing

[12] Dey, A.: Accessibility evaluation practices - survey results (2004) (Visited May 2008), http://deyalexander.com/publications/accessibility-evaluation-practices.html

[13] DRC. Formal investigation report: web accessibility. Disability Rights Commission (April 2004) (Visited January 2006), www.drc-gb.org/publicationsandreports/report.asp

[14] Gray, W.D., Salzman, M.C.: Damaged merchandise: a review of experiments that compare usability evaluation methods. Human–Computer Interaction 13(3), 203–261 (1998)

[15] Hartson, H.R., Andre, T.S., Williges, R.C.: Criteria for evaluating usability evaluation methods. Journal of Human-Computer Interaction 15(1), 145–181 (2003)

[16] Henry, S.L., Grossnickle, M.: Just Ask: Accessibility in the User-Centered Design Process. Georgia Tech Research Corporation, Atlanta, Georgia, USA, On-line book (2004), http://www.UIAccess.com/AccessUCD

[17] Hertzum, M., Jacobsen, N.E.: The evaluator effect: a chilling fact about usability evaluation methods. Int. Journal of Human-Computer Interaction 1(4), 421–443 (2001)

[18] ISO. Ergonomics of human–system interaction — guidance on accessibility for human–computer interfaces. ISO/TS 16071. Technical report, International Standards Organization (2003), www.iso.ch

[19] Italian Government. Requisiti tecnici e i diversi livelli per l'accessibilità agli strumenti informatici (July 2005) (G. U. n. 183 8/8/2005), www.pubbliaccesso.it/normative/DM080705.htm

[20] Italian Parliament. Disposizioni per favorire l'accesso dei soggetti disabili agli strumenti informatici (January 2004) (Law n. 4, January 9 2004), http://www.parlamento.it/parlam/leggi/04004l.htm

[21] Kelly, B., Sloan, D., Phipps, L., Petrie, H., Hamilton, F.: Forcing standardization or accomodating diversity? A framework for applying the WCAG in the real world. In: Harper, S., Yesilada, Y., Goble, C. (eds.) W4A 2005: Proc. of the 2005 international cross-disciplinary conference on Web accessibility, Chiba, Japan, April 2005, pp. 46–54. ACM, New York (2005)

[22] Kelly, B., Sloan, D., Brown, S., Seale, J., Petrie, H., Lauke, P., Ball, S.: Accessibility 2.0: people, policies and processes. In: W4A 2007: Proc. of the 2007 international cross-disciplinary conference on Web accessibility (W4A), pp. 138–147. ACM, New York (2007)

[23] Lang, T.: Comparing website accessibility evaluation methods and learnings from usability evaluation methods (Visited May 2008) (2003), http://www.peakusability.com.au/about-us/pdf/website_accessibility.pdf

[24] Mankoff, J., Fait, H., Tran, T.: Is your web page accessible?: a comparative study of methods for assessing web page accessibility for the blind. In: CHI 2005: Proc. of the SIGCHI conference on Human factors in computing systems, pp. 41–50. ACM, New York (2005)

[25] Nielsen, J.: Usability Engineering. Morgan Kaufmann, San Francisco (1994)

[26] Nielsen, J.: Heuristic evaluation (2002) (Visited January 2008), http://www.useit.com/papers/heuristic

[27] Nielsen Norman Group. Beyond ALT Text: Making the Web Easy to Use for Users with Disabilities (October 2001), http://www.nngroup.com/reports/accessibility/

[28] Nielsen Norman Group. Web usability for senior citizens (April 2002), http://www.nngroup.com/reports

[29] Petrie, H., Kheir, O.: The relationship between accessibility and usability of websites. In: Proc. CHI 2007, pp. 397–406. ACM, CA (2007)

[30] Petrie, H., Hamilton, F., King, N., Pavan, P.: Remote usability evaluations with disabled people. In: CHI 2006: Proc. of the SIGCHI conference on Human factors in computing systems, pp. 1133–1141. ACM, New York (2006)

[31] Rubin, J.: Handbook of usability testing. Wiley, Technical Communication Library, Chichester (1994)

[32] Sears, A.: Heuristic walkthroughs: finding the problems without the noise. Int. Journal of Human-Computer Interaction 9(3), 213–234 (1997)

[33] Slatin, J., Lewis, K.: Challenges of accessible web design: Standards, guidelines, and user testing. In: Technology and Persons with Disabilities Conference, Los Angeles, USA. CSUN, California State University Northridge (2002)

[34] Slatin, J., Rush, S.: Maximum Accessibility: Making Your Web Site More Usable for Everyone. Addison-Wesley, Reading (2003)

[35] Thatcher, J., Waddell, C., Henry, S., Swierenga, S., Urban, M., Burks, M., Regan, B., Bohman, P.: Constructing Accessible Web Sites. Glasshouse (2002)

[36] Thatcher, J., Burks, M., Heilmann, C., Henry, S., Kirkpatrick, A., Lauke, P., Lawson, B., Regan, B., Rutter, R., Urban, M., Waddell, C.: Web Accessibility: Web Standards and Regulatory Compliance (2006); Friends of ED

[37] Theofanos, M.F., Redish, J.: Bridging the gap: between accessibility and usability. Interactions 10(6), 36–51 (2003)

[38] Thevenin, D., Coutaz, J.: Plasticity of user interfaces: framework and research agenda. In: Sasse, A., Johnson, C. (eds.) Proceedings of Interact 1999, Edinburgh, UK, pp. 110–117 (1999)

[39] U.S. Dept. of Justice. Section 508 of the Rehabilitation Act (2001), www.access-board.gov/sec508/guide/1194.22.htm

[40] U.S. Government. SEC. 508. ELECTRONIC AND INFORMATION TECHNOLOGY, 1998 Amendment to Section 508 of the Rehabilitation Act, (1998) (Visited May 2008), www.section508.gov/index.cfm?FuseAction=Content&ID=14

[41] Vigo, M., Kobsa, A., Arrue, M., Abascal, J.: User-tailored web accessibility evaluations. In: HyperText 2007, Manchester, UK, September 2007, pp. 95–104. ACM, New York (2007)

[42] W3C/WAI. Conformance evaluation of web sites for accessibility (2008)(Visited May 2008), www.w3.org/WAI/eval/conformance.html

[43] W3C/WAI. How people with disabilities use the web. World Wide Web Consortium — Web Accessibility Initiative (March 2004) (Visited May 2008), http://w3.org/WAI/EO/Drafts/PWD-Use-Web/20040302.html

[44] W3C/WAI. Web accessibility evaluation tools: Overview. World Wide Web Consortium — Web Accessibility Initiative (2006) (Visited May 2008), http://www.w3.org/WAI/ER/tools/Overview.html

[45] W3C/WAI. Web content accessibility guidelines 1.0. World Wide Web Consortium — Web Accessibility Initiative (May 1999), http://www.w3.org/TR/WCAG10

[46] W3C/WAI. Web content accessibility guidelines 2.0 — w3c candidate recommendation 30 april 2008. World Wide Web Consortium — Web Accessibility Initiative (April 2008), http://www.w3.org/TR/2008/CR-WCAG20-20080430

A Systematic Review of Usability Evaluation in Web Development[*]

Emilio Insfran and Adrian Fernandez

ISSI Group, Department of Information Systems and Computation
Universidad Politécnica de Valencia
Camino de Vera, s/n, 46022, Valencia, Spain
{einsfran,afernandez}@dsic.upv.es

Abstract. The challenge of developing more usable Web applications has motivated the appearance of a number of techniques, methods and tools to address Web usability issues. Although there are many proposals for supporting the development of usable Web applications, many developers are not aware of them and many organizations do not properly apply them. This paper reports on a systematic review of the use of usability evaluation methods in Web development. The objective of the review is to investigate what usability evaluation methods have been employed by researchers to evaluate Web artifacts and how they were employed. A total of 51 research papers have been reviewed from an initial set of 410 papers. The results show that 45% of the papers reviewed reported the use of evaluation methods specifically crafted for the Web and that the most employed method is user testing. In addition, the results of the review have identified several research gaps. Specifically, 80% of the evaluations are still performed at the implementation phase of Web applications development and 47% of the papers did not present any validation of the usability evaluation method(s) employed.

Keywords: Usability Evaluation Methods, Web development, Systematic Review.

1   Introduction

Usability is a crucial factor in Web application development. The ease or difficulty that users experience with systems of this kind will determine their success or failure. As Web applications have become the backbone of business and information exchange, the need for usability evaluation methods specifically crafted for the Web – and technologies that support the usability design process – has become critical [21].

The challenge of developing more usable Web applications has motivated the appearance of a variety of techniques, methods and tools to address Web usability issues. Although there are many proposals for supporting the development of usable Web applications, many developers are not aware of them and many organizations do

---

[*] This work is funded by the META project (TIN2006-15175-C05-05), the Quality-driven model transformations project (UPV), and the CALIPSO research network (TIN2005-24055-E).

S. Hartmann et al. (Eds.): WISE 2008, LNCS 5176, pp. 81–91, 2008.
© Springer-Verlag Berlin Heidelberg 2008

not properly apply them. To address this issue, several studies aimed at comparing usability evaluation methods for Web development were reported (e.g., [1], [11]). These studies often compare a reduced number of evaluation methods, and the selection of methods is normally driven by the expectations of the researcher. Therefore, there is a need to identify, in a more systematic way, what usability evaluation methods have been successfully applied to Web development.

In this paper, we present a systematic review for assessing what usability evaluation methods have been employed for Web usability evaluation and their relation to the Web development process. Systematic reviews are useful for summarizing all existing information about a phenomenon of interest (e.g., a particular research question) in an unbiased manner [14]. The goal of our review is, therefore, to examine the current use of usability evaluation methods in Web development from the point of view of the following research questions: *what usability evaluation methods have been employed by researchers to evaluate Web artifacts and how were they employed?*

This paper is organized as follows. Section 2 discusses related work. Section 3 presents the protocol we used to review the usability evaluation methods employed in Web development. Section 4 describes the results of the systematic review. Section 5 discusses the threats to the validity of the results. Finally, section 6 presents our conclusions and suggests areas for further investigation.

2 Related Work

A number of studies aimed at comparing usability evaluation methods for Web development have been reported in the last few years (e.g., [23], [1]).

One of the most complete studies was published by Ivory and Hearst [23] in 2002. They proposed a taxonomy for classifying automated usability evaluation methods. The taxonomy was applied to 128 usability evaluation methods, where 58 of them are suitable for Web user interfaces. The results of this survey suggest promising ways to expand existing methods to better support automated usability evaluation.

Another study by Alva et al. [1] presented an evaluation of seven methods and tools for usability evaluation in software products and artifacts for the Web. The purpose of this study was to determine the degree of similarity among the methods using the principles defined in the ISO 9241-11 standard [12]. However, this is an informal survey with no defined research questions and no search process to identify the methods that were considered.

Batra and Bishu [3] reported the results obtained with two usability evaluation studies for Web applications. The objective of the first study was to compare the efficiency and effectiveness between user testing and heuristic evaluation. The results showed that both methods addressed very different usability problems and are equally efficient and effective for Web usability evaluation. The objective of the second study was to compare the performance between remote and traditional usability testing. The results indicate that there is no significant difference between the two methods.

Although several comparisons about usability evaluation methods have been reported, we are not aware of any existing systematic review published on the field of Web usability. The majority of the published studies are informal literature surveys or comparisons with no defined research questions, no search process, no defined data extraction or data analysis process. We only found two systematic reviews conducted in related fields [9], [19]. Freire et al. [9] presented a systematic review on Web accessibility to identify existing techniques for developing accessible content in Web applications. This review includes 53 studies, and it also proposes a classification of these techniques according to the processes described in the ISO/IEC 12207 standard [13]. Mendes [19] presented a systematic review to investigate the rigor of claims of Web engineering research.

3   Research Method

A systematic review is a means of evaluating and interpreting all available research that is relevant to a particular research question, topic area, or phenomenon of interest [14]. It aims at presenting a fair evaluation of a research topic by using a trustworthy, rigorous, and auditable methodology.

A systematic review involves several stages and activities. In the *planning the review* stage, the need for the review is identified, the research questions are specified, and the review protocol is defined. In the *conducting the review* stage, the primary studies are selected, the quality assessment used to include studies is defined, the data extraction and monitoring is performed, and the obtained data is synthesized. Finally, in the *reporting the review* stage, the dissemination mechanisms are specified, and the review report is presented. The activities concerning the planning and the conducting of our systematic review are described in the following subsections. The reporting the review stage is presented in Section 4.

3.1   Research Question

We have carried out a systematic literature review using the approach suggested in [14]. The goal of our study is to examine the current use of usability evaluation methods in Web development from the point of view of the following research question: *What usability evaluation methods have been employed by researchers to evaluate Web artifacts and how were they employed?* The criteria used to classify the evaluation methods are presented in Section 3.3.

This research question will allow us to summarize the current knowledge about Web usability evaluation and to identify gaps in current research in order to suggest areas for further investigation. The study's population and intervention is as follows:

- **Population:** Web usability full research papers
- **Intervention:** Usability evaluation methods
- **Outcome:** No focus on the outcome itself
- **Experimental design:** Any design

Our review is more limited than a full systematic review as suggested in [14] since we did not follow up the references in papers. In addition, we did not include other references such as technical reports, working papers and PhD theses. This strategy has been used in another systematic review conducted in the Web Engineering field [19].

3.2 Identifying and Selecting Primary Studies

The main sources we used to search for primary studies are IEEExplore and ACM digital libraries. In addition, we have included the proceedings of the following special issues and conferences:

- World Wide Web conference proceedings – WWW (2003, 2004, 2007), Usability and accessibility & Web engineering tracks [26] [7], [27].
- International conference on Web Engineering proceedings – ICWE (2003-2007) [16], [15], [17], [25], [2].
- IEEE Internet Computing Special issue on "Usability and the Web" (1 volume published in 2002) [21].
- A book on Web Engineering by Springer (LNCS) published in 2005 [20].
- International Web Usability and Accessibility workshop proceedings – IWWUA (2007) [24].

The search string defined for retrieving studies is as follows: *usability AND web AND development AND (evaluation OR experiment OR study OR testing).*

We experimented with several search strings and this one retrieved the greatest amount of relevant papers. This search string was used in the IEEExplore and the ACM digital libraries as well as in the other sources that were inspected manually. The period reviewed was the last 10 years, i.e., studies published from 1998 to 2008. With respect to the digital libraries, we ensured that our search strategy was applied to magazines, journals and conference proceedings.

3.3 Inclusion Criteria and Procedures

Each identified study was evaluated the researchers conducting the systematic review to decide whether or not it should be included. The discrepancies were solved by consensus. The studies that met the following conditions were included:

- Studies presenting usability evaluation method(s) that are applied to Web development. Only studies that presented a "formal" method (e.g., heuristic evaluation, cognitive walkthrough) were selected.
- Full research papers.

The following types of papers were excluded:

- Papers presenting recommendations and principles for Web design.
- Papers presenting techniques on how to aggregate usability measures.
- Papers presenting usability metrics.
- Introductory papers for special issues, books, and workshops.
- Papers not written in English.

3.4 Data Extraction Strategy

The data extracted were compared according to the research questions stated, which are decomposed into the following criteria:

1. What usability evaluation methods (UEMs) have been employed by researchers to evaluate Web artifacts?

      i.   Is it a new evaluation method or an existing method from the HCI field? (New, Existing)

      ii.   What is the type of usability evaluation method employed? (Inspection method, User testing, Other)

2. What is the phase in which the evaluation method is applied? (Requirements, Design, Implementation)

3. What is the type of evaluation? (Manual, Automated)

4. Was the evaluation method evaluated? (Yes, No). If yes:

      i.   What type of evaluation was conducted? (Survey, Case study, Experiment)

5. Was the evaluation conducted with the intention to provide feedback to the design? (Yes, No)

With regard to the first criterion, the paper is classified as *new* if it presents at least one evaluation method that is specifically crafted for the Web. Otherwise, it is classified as *existing* if the paper uses existing methods from the HCI field.

In addition, the evaluation method is classified according to the following types: inspection method, user testing, or other. The paper is classified as *inspection method* if it reports an evaluation based on expert opinion (e.g., heuristic evaluation, guideline reviews, standards inspection, cognitive walkthroughs). Otherwise, the paper is classified as *user testing* if it reports an evaluation that involves the user's participation. Such evaluations typically focus on lower-level cognitive or perceptual tasks. In this category, we also consider the several protocols that exist to conduct user testing (e.g., thinking aloud, question-asking). Finally, the paper is classified as *others* if it reports the use of other methods (e.g., focus group, web usage analysis).

With regard to the second criterion (the phase in which the evaluation is conducted), each paper is classified into one or more ISO/IEC 12207 high-level processes: Requirements, Design, and Software Construction (Implementation). The paper is classified at the *requirements* phase if the artifacts used as input for the evaluation include high-level specifications of the Web application (e.g., task models, uses cases, scenarios). The paper is classified at the *design* phase if the evaluation is conducted on the intermediate artifacts of the Web application (e.g., navigational models, abstract user interface models, dialog models). Finally, the paper is classified at the *implementation* phase if the evaluation is conducted in the Web application.

With regard to the third (the type of evaluation conducted), the paper is classified as *manual* if it presents a usability evaluation that is manually performed. Otherwise, it is classified as *automated*. The fourth criterion is related to the evaluation of the usability evaluation methods. Depending on the purpose of the evaluation and the conditions for empirical investigation, three different types of strategies can be carried out [8]: survey, case study and experiment. A *survey* is an investigation performed in retrospect, when the method has been in use for a certain period of time. A *case study* is an observational study and data is collected for a specific purpose throughout the study. An *experiment* is a formal, rigorous and controlled investigation. Experiments provide a high level of control and are useful for comparing usability evaluation methods in a more rigorous way. For evaluations of this type, statistical methods are applied in order to determine which method is better.

Finally, the fifth criterion is to determine whether or not the evaluation method provides feedback to the designer. The evaluation method is classified as *No* if it is aimed at only reporting usability problems. The method is classified as *Yes* if it also provides recommendations on how the problems can be fixed.

3.5 Conducting the Review

The search to identify primary studies in the IEEEplore and ACM digital libraries was conducted on the 22nd of March 2008. The application of the review protocol yielded the following results:

- The bibliographic database search identified 338 potentially relevant publications (181 from the IEEEplore and 157 from the ACM digital library). After applying the exclusion criteria documented in Section 3.3, 37 publications were finally selected (11 from IEEEplore and 26 from ACM digital library).
- The manual bibliographic review of the other sources identified another 72 potentially relevant publications. After applying the exclusion criteria, the following publications were finally selected: 14 papers (3 from WWW, 3 from ICWE, 3 from the IEEE Internet Computing special issue, 4 from IWWUA, and 1 chapter from the book).

Therefore, a total of 51 research papers were selected by our inclusion criteria. Some studies had been published in more than one journal/conference. In this case, we selected only the most complete version of the study. Other studies appeared in more than one source. These publications were taken into account only once. The searching results revealed that research papers about Web usability are published in several conferences/journals from different fields, such as Human-Computer Interaction (HCI), Web Engineering (WE), and other related fields.

4 Results

The results of our study are presented in Table 1. They have been organized by selection criteria and publication source. The list of papers containing all the data extracted from the studies was not included in this paper due to space restrictions.

These results indicate that 45% of the papers reviewed presented new evaluation methods specifically designed for the Web (see Fig. 1 (a)). For instance, Blackmon et al. [5] proposed the cognitive walkthrough for the web (CWW) method. When compared to the traditional method, this method was found to be superior for evaluating how well websites support user navigation and information search tasks. In another study, Bolchini and Garzotto [6] proposed a usability inspection method for Web applications called MiLE+. The method was evaluated through two studies that measured the efficiency, performance, and the perceived difficulty of learning the method. The remaining 55% of the studies reported the use of existing evaluation methods (e.g., cognitive walkthrough, heuristic evaluation, user testing).

The results also revealed that the most frequently used type of evaluation method is user testing, i.e., 41% of the papers reviewed reported some kind of testing involving users (see Fig. 1 (b)). This may indicate that most evaluations are performed mainly

Table 1. Systematic Review Results

Selection criteria		IEEE	ACM	WWW	ICWE	IE3IC	Book	IWWUA	Total
Usability	New	4	9	2	3	3	0	2	23
Evaluation Method	Existing	7	17	1	0	0	1	2	28
Type of Usability	Inspection method	4	5	0	1	1	1	1	13
Evaluation Method	User testing	7	17	1	0	0	1	0	26
	Other	4	11	2	2	2	1	3	25
Web devel-opment phase	Requirements	1	1	0	0	0	0	1	3
	Design	4	4	0	1	3	1	3	16
	Implementation	7	25	3	3	1	1	1	41
Type of evaluation	Manual	9	19	0	1	1	1	4	35
	Automated	2	7	3	2	2	0	0	16
Validation?	Survey	0	3	0	0	0	0	0	3
	Case study	1	3	2	1	0	1	3	11
	Experiment	2	10	0	0	0	0	1	13
	No	8	10	1	2	3	0	0	24
Feedback to	Yes	4	6	0	0	2	0	3	15
design?	No	7	20	3	3	1	1	1	36

IEEE – IEEExplore electronic database
ACM – ACM digital library
WWW – World-Wide Web conference from 2003 to 2007
ICWE – International Conference on Web Engineering from 2003 to 2007

IE3IC – IEEE Internet Computing Special Issue on Usability and the Web
Book – A book on Web Engineering by Springer
IWWUA – International Workshop on Web Usability and Accessibility 2007

during late stages of the Web development lifecycle. Inspections accounts for 20% of the studies, whereas 39% of the studies reported the use of other methods (e.g., paper prototype, remote user testing, survey). An example of the use of inspection methods is described in Sutcliffe [22]. The author proposed a set of heuristics for assessing the attractiveness of Web user interfaces. The heuristics were tested by evaluating three airline websites. The results of the study show that aesthetics may play an important role for initial visits but content issues may be dominant for repeat visits.

The analysis of the results confirmed that the evaluations are mainly performed at the implementation level (68%) of the Web application (see Fig. 1(c)). Around 27% of the studies describe evaluations performed using the Web application's intermediate artifacts (e.g., abstract user interface, navigational model). Only 5% of the evaluations were performed at the requirements specification level (e.g., laboratory user testing of paper mock-ups or prototypes). Therefore, there is a need for usability evaluation methods that can be used at early stages of Web development.

With regard to the type of evaluation, 69% of the studies performed the evaluations manually (see Fig. 1 (d)). Around 31% of the studies reported the existence of some kind of automated tool to support the proposed method. For instance, Becker and Berkemeyer [4] proposed a technique to support the development of usable Web applications. The technique is supported by a GUI-based toolset called RAD-T (rapid application design and testing) that allows early usability testing at the design stage.

We also verified whether the studies reported some kind of empirical evaluation. The results revealed that 47% of the studies did not conduct any type of evaluation (see Fig. 1 (e)). However, it was surprising to observe that, from the papers that did perform evaluations, 25% of them reported on controlled experiments. The majority of these studies were published in HCI conferences and journals; hence, experimentation is a common research method used in this field. An example of this is the study conducted by Hornbæk and Frøkjær [11], where two psychology-based inspection techniques (cognitive walkthrough (CW) and metaphors of human thinking (MOT)) were compared. The results show that the participants identified 30% more usability problems using MOT. Around 22% of the studies report case studies. For instance, Matera et al. [18] presented a case study in which three methods were applied to the evaluation of a Web application: design inspections to examine the hypertext specification, web usage analysis to analyze the user behavior, and heuristic evaluation to analyze the released prototypes and the final Web application.

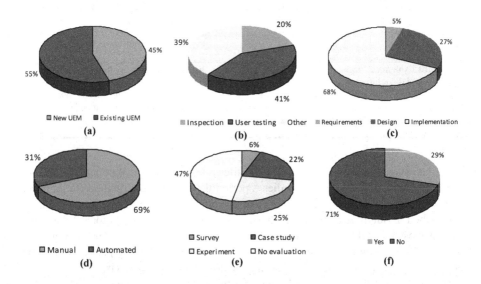

Fig. 1. Percentage of coverage by criteria used for data extraction

Finally, 71% of the studies reported only on usability problems giving no feedback on the corresponding design artifacts (see Fig. 1 (f)). The remaining studies (29%) also offered suggestions for design changes based on the usability problems detected. For instance, Hornbæk and Frøkjær [10] reported an experiment aimed at comparing the assessment of both usability and utility of problems and redesign suggestions. The results of the experiment showed how redesign proposals were assessed by developers as being of higher utility than just problem descriptions. Usability problems were seen more as a help in prioritizing ongoing design decisions.

Figure 2 shows the number of selected publications on Web usability evaluation methods by year and source. The analysis of the number of research studies on Web usability showed that there has been a growth of interest on this topic. Most of the studies about Web usability were found at the ACM digital library.

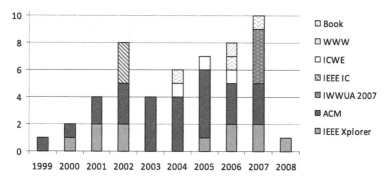

Fig. 2. Number of Publications on Web Usability by Year and Source

5 Threats to Validity

The main limitations of this study are publication selection bias, inaccuracy in data extraction, and misclassification. Publication bias refers to the problem that positive results are more likely to be published than negative results [14]. We believe that we have alleviated this threat, at least to some extent, by scanning relevant journal special issues and conference proceedings. However, we did not consider grey literature or unpublished results. With regard to publication selection, we chose the sources where papers about Web usability are normally published. However, we have excluded some journals in the Web engineering field from this systematic review (i.e., Journal of Web Engineering and International Journal of Web Engineering and Technology) since we had no access to these journals. This fact could affect the validity of our results. We attempted to alleviate the threats of inaccuracy in data extraction and misclassification by conducting the classifications of the papers with three reviewers.

6 Conclusions and Future Work

This paper has presented a systematic review of usability evaluation methods for Web development. The results of the review have identified several research gaps.

In particular, usability evaluations should be performed early in the Web development process and should occur repeatedly throughout the design cycle, not just when the product has been completed. The majority of the papers reported on evaluations at the implementation phase. It also reveals that the evaluations are mainly performed in a single phase of the Web application development. Usability evaluation at each phase of the Web application development is critical for ensuring that the product will actually be used and be effective for its intended purpose(s). In addition, the majority of the methods reviewed only allowed the generation of a list of usability problems. New proposals for redesign that address usability problems as an integral part of the evaluation method are needed.

Although our findings may be indicative of the field, further reviews are needed to confirm the results obtained. Future work includes the extension of this review by including other sources (e.g., Science Direct and Scopus databases). We also want to analyze more in-depth the level of integration of the usability evaluation methods into the different processes of the Web application lifecycle. Finally, we plan to collect more information about the empirical evidence of the effectiveness of usability evaluation methods for the Web.

References

1. Alva, O.M.E., Martínez Prieto, A.B., Cueva Lovelle, J.M., Sagástegui Ch., T.H., López, B.: Comparison of Methods and Existing Tools for the Measurement of Usability in the Web. In: Proc. Int. Conf. on Web Engineering, Spain, pp. 386–389. Springer, Heidelberg (2003)
2. Baresi, L., Fraternali, P., Houben, G.-J. (eds.): ICWE 2007. LNCS, vol. 4607. Springer, Heidelberg (2007)
3. Batra, S., Bishu, R.R.: Web Usability and Evaluation: Issues and Concerns. In: Aykin, N. (ed.) HCII 2007. LNCS, vol. 4559, pp. 243–249. Springer, Heidelberg (2007)
4. Becker, S.A., Berkemeyer, A.: Rapid Application Design and Testing of Web Usability. IEEE Multimedia 9(4), 38–46 (2002)
5. Blackmon, M.H., Polson, P.G., Kitajima, M., Lewis, C.: Cognitive walkthrough for the web. In: Proc. of the CHI 2002, Minneapolis, Minnesota, USA, pp. 463–470 (2002)
6. Bolchini, D., Garzotto, F.: Quality of Web Usability Evaluation Methods: An Empirical Study on MiLE+. In: Proc. of the IWWUA 2007, Nancy, France, pp. 481–492 (2007)
7. Feldman, S.I., Uretsky, M., Najork, M., Wills, C.E.: Proc. of the International Conference on World Wide Web 2004, pp. 17–20. ACM, New York (2004)
8. Fenton, N., Pfleeger, S.L.: Software Metrics: A Rigorous and Practical Approach, 2nd edn. International Thomson Computer Press (1996)
9. Freire, A.P., Goularte, R., Fortes, R.P.M.: Techniques for Developing more Accessible Applications: a Survey Towards a Process Classifications. In: Proc. of the 25th ACM Int. Conference on Design of communication, El Paso, Texas, USA, pp. 162–169 (2007)
10. Hornbæk, K., Frøkjær, E.: Comparing Usability Problems and Redesign Proposals as Input to Practical Systems Development. In: Proc. of the CHI 2005, Portland, USA, pp. 391–400 (2005)
11. Hornbæk, K., Frøkjær, E.: Two psychology-based usability inspection techniques studied in a diary experiment. In: Proc. of the 3rd Nordic conference on Human-computer interaction (NordCHI 2004), Tampere, Finland, pp. 3–12 (2004)
12. ISO – International Standard Organization, ISO 9241-11: Ergonomic requirements for office work with visual display terminals (VDTs) – Part 11: Guidance on usability (1998)
13. ISO – International Standard Organization, ISO/IEC 12207: Standard for Information Technology – Software Lifecycle Processes (1998)
14. Kitchenham, B.: Guidelines for Performing Systematic Literature Reviews in Software Engineering. Version 2.3, EBSE Technical Report, Keele University, UK
15. Koch, N., Fraternali, P., Wirsing, M. (eds.): ICWE 2004. LNCS, vol. 3140. Springer, Heidelberg (2004)
16. Cueva Lovelle, J.M., Rodríguez, B.M.G., Gayo, J.E.L., Ruiz, M.d.P.P., Aguilar, L.J. (eds.): ICWE 2003. LNCS, vol. 2722. Springer, Heidelberg (2003)
17. Lowe, D.G., Gaedke, M. (eds.): ICWE 2005. LNCS, vol. 3579. Springer, Heidelberg (2005)
18. Matera, M., Rizzo, F., Carughi, G.T.: Web Usability: Principles and Evaluation Methods. In: Mendes, E., Mosley, N. (eds.) Web Engineering, pp. 143–179.

19. Mendes, E.: A Systematic Review of Web Engineering Research. In: Proc. of the International Symposium on Empirical Software Engineering (ISESE 2005), pp. 498–507 (2005)
20. Mendes, E., Mosley, N. (eds.): Web Engineering. Springer, Heidelberg (2005)
21. Neuwirth, C.M., Regli, S.H.: IEEE Internet Computing Special Issue on Usability and the Web. IEEE Internet Computing Special Issue on Usability and the Web 6(2) (March/April 2002)
22. Sutcliffe, A.: Assessing the Reliability of Heuristic Evaluation for Website Attractiveness and Usability. In: Proc. of the HICSS 2002, vol. 5, pp. 137–141.
23. Yvory, M., Hearst, M.: The State of the Art in Automating Usability Evaluation of User Interfaces. ACM Computing Surveys 33(4), 470–516 (2001)
24. Weske, M., Hacid, M.S., Godart, C. (eds.): Web Information Systems Engineering - WISE 2007 Workshops Proceedings, Nancy, France, December 3, 2007. LNCS, vol. 4832. Springer, Heidelberg (2007)
25. Wolber, D., Calder, N., Brooks, C.H., Ginige, A. (eds.): Proc. of the 6th International Conference on Web Engineering 2006, alo Alto-CA, USA, July 11-14. ACM, New York (2006)
26. WWW 2003, Proc. of the Twelfth International World Wide Web Conference 2003, Budapest, Hungary, May 20-24, 2003. ACM, New York (2003), http://www2003.org
27. WWW 2007, Proc. of the Sixteenth International World Wide Web Conference, Banff, Alberta, Canada, May 8-12, 2007. ACM, New York (2007), http://www2007.org

Identifying Semantic Constructs in Web Documents to Improve Web Site Accessibility

Mathias Koehnke[1], Temenushka Ignatova[1], Martina Weicht[2], and Ilvio Bruder[2]

[1] Dept. of Computer Science, University of Rostock, Germany
mathias.koehnke@gmail.com, temi@informatik.uni-rostock.de
[2] IT Science Center Ruegen gGmbH, Germany
{weicht,bruder}@it-science-center.de

Abstract. Misleading user interfaces and overloaded web sites are some of the reasons why users avoid certain web sites while searching for information on the world wide web. In order to improve the usability and accessibility of web sites, techniques which take the semantic structure of the web documents into account have to be employed. The semantic analysis approach described in this paper aims at recognizing those parts within web documents that are particularly relevant for a specific usage scenario. Different combinations of syntactical constructs are mapped to different analysis classes whose semantic constructs correspond to concrete interaction tasks predefined in task models. The task models are used as a workflow guiding the user through the shopping procedure, whereby for each identified task a corresponding semantic concept on the web site is identified. The described techniques will be embedded in a so called screen reader application for visually impaired people who do not have the ability to use a graphical display.

1 Introduction

Usability and accessibility are two of the main requirements of the general user browsing the web. Especially disabled people, e.g. the visually impaired, have significant problems with the current web of misleading user interfaces and overloaded web sites. Often, a simple presentation such as WAP sites (Wireless Application Protocol) offer more user-friendly interfaces than their full-fledged WWW pendant. For example, visually impaired people use the WAP version of Amazon for ordering audio books, because the WWW version of the site has no clearly recognizable structure for important tasks such as search, browse, and purchase.

We introduce an analysis approach for finding a semantic structure within a web site. Both the structure and its semantics are very helpful when trying to understand the intended usage of a web site and to find the right information therein. Web sites are built from syntactical constructs forming semantic constructs which correspond to certain usage and interaction tasks. Discovering these semantic constructs enables the user to better exploit, navigate and obtain information from the web site. Syntactical constructs, however, can have

S. Hartmann et al. (Eds.): WISE 2008, LNCS 5176, pp. 92–101, 2008.

ambiguous meanings making the automatic recognition of semantic constructs a challenging task. In order to solve this problem, we use task models to describe a web site's semantic domain. The tasks within a task model form a hierarchical set of semantic descriptions for interrelated interaction activities: They are arranged in a specific order representing the interaction process. The analysis approach presented in this paper matches the semantic description of a task with structure elements and interaction possibilities within a web site. The results of this matching depend largely on the quality of the task model as well as on the evaluation technique used for the matching. For our evaluation we employ data mining techniques, namely neural networks.

For this analysis approach we implemented a prototype using the Unstructured Information Management Architecture (UIMA)[1] developed by IBM. The motivation and requirements for this development come from the project SUE (Screenreader Usability Extensions) for visually impaired people which is funded by the German government. While contemporary screen readers simply read out the text from the graphical display, the introduced analysis technique suggests an improvement for the screen reader technology by guiding the user through the interaction tasks.

2 Related Work

Related techniques for solving the problem with identifying useful information in web documents have been developed mainly in the areas of web mining [3,8], web accessibility and user interface design. Web mining techniques try to recognize and extract data from web-based sources.

Many approaches in the field of discovering semantic structure in documents deal with partitioning (web) documents into structural blocks, either by analyzing the document image [1], by employing wrapper learning [2] or by evaluating the DOM representation of the original HTML source code [4,10]. These, however, do not take into account any specific structures they are looking for, but try to divide the web site into certain areas of interest or classify them according to certain criteria. In our approach, the structuring follows specific tasks to be performed. Those tasks are either represented by a task model describing a complex task via its subtasks or belong to a set of pre-defined structural constructs that can be found on a number of web sites. This guidance leads to more specific results when trying to find certain elements on complex web sites, but on the other hand, requires some preliminary work preparing task models and descriptions of structural constructs. Those are, however, of a general nature allowing them to be reused on a number of web sites and therefore outweighing the extra work as soon as a basic set of descriptions and task models is defined.

In the area of web accessibility there are interesting approaches for tools and technologies improving the accessibility of structured web sites. The Dante approach of Yesilada, Stevens, and Goble [13] introduces a tool for identifying and classifying travel objects, meaning interaction objects, in web sites based

---

[1] http://uima-framework.sourceforge.net/

on heuristics. Discovering their inherent semantic is based on a very similar approach. In [11], the Dante Tool is extended by semantic annotation concepts using ontologies. This extension provides an automatic technique to annotate web sites considering their design, i.e. the authoring process. It seems difficult though to involve the design process when using a web site if the web creator is not available. In our approach, we want to tackle the annotation problem by including the existing helping and scripting community along with a semi-automatic approach. Harper and Bechhofer [6] describe concepts for structural semantics for accessibility and device independence. They, however, rely on CSS which, unfortunately, are note used in every web site. In contrast, all web sites do contain HTML elements, therefore this seems to be a more general approach.

The concept of task models is part of model-based software development and is mainly used for designing user interfaces [9,12]. All tasks including user interaction and system reaction are modeled and expressed in a task tree.

3 Descriptors for Semantic Constructs of Web Documents

In this paper, semantic constructs are the basic concept used for describing the semantic structure of web documents. Navigation menus, login fields, search fields, shopping cart, terms and conditions, help, and site map are examples for such constructs. Semantic constructs represent building blocks of a web document which correspond to a particular interaction activity of the web site's usage scenario. This usage scenario is defined as a task model which guides the user through the different tasks in the right order. Fig. 1 shows an exemplary task model for "Online Shopping".

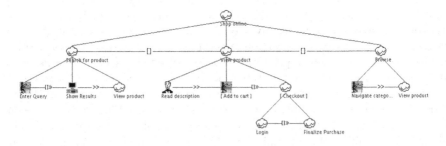

Fig. 1. Sample task model "Online Shopping"

The complex task "Search for product" requires the user to enter her query, i.e. the name of an author or the title of a CD. Using a screen reader, finding the search field takes quite some time - especially on an unknown web site - or even fails. Therefore, when reaching the task "Enter Query" the cursor could automatically be positioned within the right text entry field. This is achieved by mapping each semantic construct to a combination of syntactical elements. These combinations are derived from the analysis of a sufficient amount of web sites containing the described semantic constructs. In our example, the search field consists of a text field, a label and a button, but, of course, other combinations may occur within other web sites.

In a preparatory step of our approach, each task of the task model is mapped to a semantic construct which can be found on web sites. The mapping is stored in a so called mapping table. During the navigation through a particular web site, for a given task the concrete semantic constructs are identified and their position in the web document is returned to the user. This step includes evaluating the description of a semantic concept against the navigated web page and making a decision about whether the semantic concept is found or not. Fig. 2 shows the desired result of applying these two steps to our sample usage scenario.

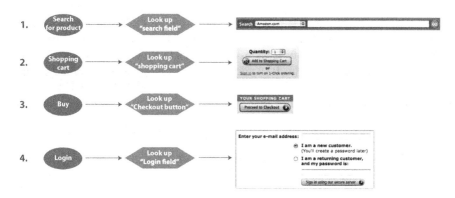

Fig. 2. Desired mapping of "Online Shopping" tasks to structural constructs of the Amazon web site

Semantic constructs of a web document mostly comprise more than one component. These components can be other semantic constructs or directly identifiable syntactical elements of a web document, such as a text field, a caption, a button or a drop-down list control. In this paper, such elements are referred to as *syntactical elements*. The characteristics or *features* of a semantic construct, which can be used to identify it, are thus the types of its components, their relative and absolute position and their content. Each semantic construct, i.e. its *features*, is described as a set of Resource Description Framework (RDF) expressions. These expressions comprise of a resource (subject), an aspect (predicate) and a statement (object). They describe the existence of a *feature* of a *syntactical element* within a web document. The RDF schema, used to build the descriptions of semantic constructs, comprises properties and functions corresponding to all *syntactical elements*, which may be found in an HTML document. It is used as a dictionary for the structural analysis. The RDF schema is used to generate an RDF document (instance) for each semantic construct. This task should be performed by an expert web site developer by analyzing a sufficient amount of web sites containing the semantic construct and deriving a generic description

of this construct in terms of *syntactical elements*. An instance of each syntactical element relevant for the given semantic construct and its corresponding
properties is included in the RDF document. The syntax of the RDF documents
enables also the definition of hierarchies of syntactical elements as well as alternative syntactical elements. These possibilities are very useful because different
web sites use different combinations of syntactical elements to represent the
same semantic construct. An example of the graphical representation of an RDF
document, describing the semantic construct "search field" is shown in Fig. 3.
A "search field" is a semantic construct frequently found in various shopping,
community, and blog sites. By analyzing the source code of web document parts
containing a "search field", it can be concluded that this semantic construct is
most often surrounded by a <form> element (#**SearchForm**). All other search
elements are embedded in this search form, therefore, the RDF description of the
search field has a hierarchical structure. At least two of the embedded elements,
a search text field (#**SearchInputText**) and a search button, which can have
different forms (#**SearchInputSubmit, #SearchInputImage** etc.), must be
present in the form to identify it as a "search field". This mandatory requirement is also represented in the RDF document through the set type **rdf:Bag**.
The different alternatives for the search buttons are modeled in RDF with the
help of the **rdf:Alt** type.

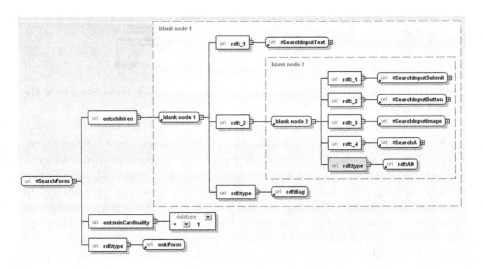

Fig. 3. A graphical representation of the RDF schema for the semantic construct
"search field"

4   Identification of Semantic Constructs

For each task of the task model a corresponding semantic construct has to be
identified on the web page. The first step towards this identification is to evaluate the RDF statements which represent the description of a semantic construct

against the DOM tree representation of the web document. The evaluated statements are used in a second step for classifying the syntactical elements found as a semantic concept, i.e. for deciding whether the semantic concept is present on the web page. The whole analysis process is illustrated in Fig. 4.

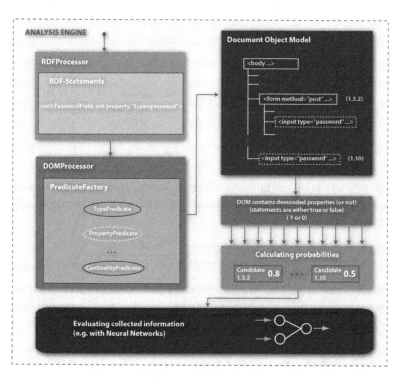

Fig. 4. The semantic structure analysis process

Evaluation of the Descriptors. The RDF statements are interpreted as predicates which are evaluated to a true or false value depending on whether the *feature* was found or not found in the document. The order in which these statements are evaluated is predefined for each construct. In Fig. 5, a statement tree determining the evaluation order for a login field construct is illustrated. Three types of statements can be distinguished in the evaluation process: type statements, literal statements and complex statements. Type statements are declared with an **rdfs:type** predicate and represent the existence of an HTML syntactical element in the semantic construct. The type statements are directly evaluated to true or false depending on whether the syntactical element is present or not present in the web document. If a syntactical element is not present its child (literal) statements and in the case of a complex statement its ancestors do not have to be evaluated. The literal statements describe the properties of a syntactical element. Complex statements contain a Uniform Resource Identifier (URI) pointing to a child statement which needs to be processed recursively.

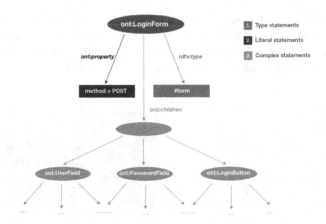

Fig. 5. Evaluation order for the RDF statements of a "login field" construct

Each RDF statement might cover different sections of the DOM tree, called candidates. For each of these candidates a *Probability Value* is calculated using the following formula.

$$Probability\ Value = \sum \frac{Truth\ Value}{Number\ of\ Attributes} \qquad (1)$$

The *Truth Values* of all the evaluated literal statements describing a syntactical element is divided by the number of these attributes. The candidate with the highest *Probability Value* is used for further evaluation of the semantic construct.

Interpretation of the Extracted Information. Once the RDF statements are evaluated, i.e. their *Probability Values* have been assigned, the interpretation of these values can begin. In this step, we can use an empirically determined threshold value over all *Probability Values* to decide if the semantic concept we are looking for is present in the document. The problem with this approach is that the different RDF statements are of different importance for this decision. Therefore, we chose to apply classification methods which can calculate the weights of the different RDF statements in a training phase, before deciding whether the candidate corresponds to a given semantic construct or not.

In this paper, artificial neural network are used as a decision making mechanism. The *Probability Values* of the syntactical constructs, represented by the RDF statements are used as input values for the neural network. The neural network calculates the probability that the requested semantic construct was found on the web page. If the semantic construct is found, its position data is sent to the screen reader application as shown in Fig. 6. The neural network in Fig. 6 determines the probability that a login field is found on the web page, using the input data from the evaluated RDF document and the weights of the network nodes, which are previously calculated with the help of training data. If the determined *Probability Value* is higher than a certain threshold, the DOM document part is identified as the desired semantic construct. In the training phase for each semantic construct a neural network has to be created with the help of so called positive and negative "URL lists". These lists contain URLs of web sites where the semantic construct is present and web sites which do not contain this construct, respectively. The RDF documents corresponding to the semantic constructs have to be evaluated for each of these URLs. The results are then used as training data for the neural network. The advantage of the neural network approach is that it performs very well for fuzzy data. However, a well known disadvantage of this classification method is the problem with overfitting.

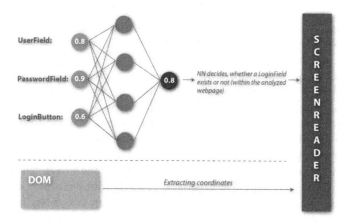

Fig. 6. Interpretation of the Descriptors with the artificial neural network

5   Implementation and Prototype Testing

In order to provide a high degree of flexibility of the semantic structure analysis, an implementation of the developed concept was carried out using the UIMA Framework. UIMA is a platform-independent Java framework for creating applications which analyze unstructured information and extract relevant knowledge out of them [5]. Any new analysis software based on UIMA is based on an algorithm fulfilling the desired task. All input and output parameters are defined by the component interface. With the help of a Component Description the component containing the algorithm is integrated into UIMA which in turn creates a UIM application out of this specification. For the implementation of the semantic structure analysis technique three basic components of the UIMA framework were used, namely, the Collection Reader, the Analysis Engine, and the CAS Consumer [7]. The semantic structure analysis is implemented in Java and comprises different steps. If the analysis will be guided by a task model (in our case the online shopping model), all tasks within this model are analyzed and an order of execution is determined. The task model is described in an XML document available either in the local file system or via a URL on the web. One way to describe a task model is to create a graphical representation using CTTE, the Concurrent Task Tree Environment [9] and use its included export to XML. An internal mapping table contains information about which RDF document is needed to provide the description of the semantic construct connected to a certain task within the task model. Selected RDF documents are read by the RdfFileParser. If no task model will be used, all RDF documents are read. They are processed in the determined order, delivering probability values based on the DOM and determined by the analysis classes. From the set of all probability values, a neural network computes an average probability value which needs to exceed a certain threshold to declare that the source document contains the desired semantic structure. In this case, its position data is displayed and passed to the screen reader.

The prototypical implementation of the structural analysis introduced in this paper has been tested with about 20 randomly chosen news and community web sites containing the semantic construct *search field*. In all of them, the semantic construct was found and in 75% of all cases, its position was correctly determined. With an average processing time of less than one second per construct the algorithm produces an acceptable result that meets the requirement for adequate reaction time in screen reader applications. It remains yet to be seen why positions are not always correctly identified. One reason can be found in invalid HTML documents possibly requiring some kind of pre-processing and normalisation. Also, the algorithm tends to fail with more than one search field construct within one single web site, i.e. web sites permitting different kind of searches. In this case the construct achieving a higher rating is favoured regardless of the search field's actual purpose. Further testing of other sematic constructs is required to ensure that a) their RDF descriptions are adequate, b) the algorithm works as reliably as presumed, and c) the proposed structural analysis can support a user performing a complex task on an overloaded web site. While the test results at hand merely provide a proof of concept, more comprehensive tests will ensure performance and quality once this approach is integrated into the SUE project and its screen reader.

6   Conclusions

Client side analysis techniques can significantly improve web site accessibility and usability and thus provide support especially to visually impaired people. In this paper, we propose an automatic recognition of the meaning of interaction elements in web sites. These techniques comprise constraining the semantic domains using task models, matching syntactical to semantic constructs as well as weighting the reliability of recognized semantic constructs. Our solution offers an alternative web site navigation using predefined tasks. It does not intend to set new limits or barriers for the users. For a sustainable usage, it is important to extend and stimulate the existing scripting and helping community who adds and updates task models and RDF descriptions. It also remains to be tested, and, if necessary, adopted to dynamic web sites.

References

1. Berardi, M., Lapi, M., Malerba, D.: An Integrated Approach for Automatic Semantic Structure Extraction in Document Images. Document Analysis Systems, Florence (2004)
2. Cohen, W.W.: Learning and Discovering Structure in Web Pages IEEE Data Engineering Bulletin. 26 (2003)
3. Etzioni, O.: The world wide web: Quagmire or gold mine. Communications of the ACM 39(11) (1996)
4. Feng, J., Haffner, P., Gilbert, M.: A Learning Approach to Discovering Web Page Semantic Structures. In: Proc. of the 8th Conference on Document Analysis and Recognition, Seoul (2005)
5. Ferrucci et al.: IBM Research Report. Towards an Interoperability Standard for Text and Multi-Model Analytics. Technical Report. IBM (2006)
6. Harper, S., Bechhofer, S.: SADIe: Structural Semantics for Accessibility and Device Independence. ACM Transactions on Computer-Human Interaction 14(2) (2007)
7. Koehnke, M.: Recognition of Structure and Semantic Relationships in Web Documents. University of Rostock. Diploma Thesis (in German) (2008)
8. Kosala, R., Blockeel, H.: Web Mining Research: A Survey. In: SIGKDD Explorations. ACM SIGKDD (July 2000)
9. Mori, G., Paterno, F., Santoro, C.: CTTE: Support for Developing and Analyzing Task Models for Interactive System Design. IEEE Transactions on Software Engineering 28(8) (2002)
10. Mukherjee, S., Yang, G., Tan, W., Ramakrishnan, I.V.: Automatic Discovery of Semantic Structures in HTML Documents. In: Proceedings of the 7th International Conference on Document Analysis and Recognition (ICDAR 2003), Edinburgh (2003)
11. Plessers, P., Casteleyn, S., Yesilada, Y., De Troyer, O., Stevens, R., Harper, S., Goble, C.: Accesibility: A Web Engineering Approach. In: Proc. of the Int. World Wide Web Conference (2005)
12. Radeke, F., Forbrig, P.: Patterns in Task-Based Modeling of User Interfaces In: Task Models and Diagrams for User Interface Design (2007)
13. Yesilada, Y., Stevens, R., Goble, C.: A foundation for tool based mobility support for visually impaired Web users. In: Proc. of the 12. World Wide Web Conf. (2003)

Improving the Usability of Novel Web Software: An Industrial Case Study of an Institutional Repository

Dana McKay and Shaun Burriss

Information Resources, Swinburne University of Technology,
John Street, Hawthorn, Vic 3123 Australia
{dmckay,sburriss}@swin.ed.au

Abstract. In this paper we discuss usability improvements made to institutional repository software in an industrial setting. The novelty of institutional repository software as a class and the industrial setting present special challenges to traditional usability evaluation; here examine these problems and present a case study which demonstrates a combination of heuristic analysis, empirical methods, background research and interface best practice to make significant usability improvements to an institutional repository.

Keywords: Institutional repositories, usability, case study, search interfaces, heuristic analysis.

1 Introduction

Institutional repositories, while a relatively new concept, are becoming widely implemented, largely due to institutional and funder mandates for self archiving [1], dependence of research assessment on repositories [2], and the rise of the open access movement [3].

However, like many new technologies, there has been little interest in the usability of institutional repositories. Research has shown that repositories will stand near-empty and ignored if there is no investigation into how they work for their contributors [4, 5]. End users of repositories fare even worse than contributors, yet at the time of writing we could find only a single published usability study of any institutional repository from an end-user standpoint [6], and the scope of this work is limited to comparing two repository systems with each other without any wider assessment of usability. This dearth of work is alarming, because it is well demonstrated that users quickly give up on information systems they cannot use [7-9], either by leaving without the information they needed [10], or by "satisificing" [11]. Institutional repositories must be usable if they are to be used at all.

In this paper we present an industrial-context case study on improving an institutional repository from an end-user standpoint. We specifically address the difficulties in evaluating a new type of system in an industrial setting, and we discuss the improvements made as a result of the evaluation. The repository in question is Swinburne University of Technology's institutional repository, Swinburne Research Bank, which is a VTLS VITAL repository with over 7000 records, including about 14 percent full text at the time of writing. This repository is managed by Swinburne Library.

S. Hartmann et al. (Eds.): WISE 2008, LNCS 5176, pp. 102–111, 2008.

In Section 2 we discuss the background context of this work and the methodology we selected; in Section 3 we review some of the background research that informed our evaluation; in Section 4 we demonstrate some example improvements to the Swinburne Research Bank interface, and in Section 5 we draw conclusions and in Section 6 we discuss necessary future work in the field of institutional repository usability.

2 Background and Methodology

The methodology we chose to evaluate the user interface of our institutional repository was primarily based on heuristic analysis. We give some of the background to this choice in Section 2.1, and describe our approach more fully in Section 2.2.

2.1 Background

There were a number of factors affecting the choices made about usability methodology when improving Swinburne's institutional repository, including the lack of background understanding of institutional repositories, the industrial context in which this evaluation took place, and the development cycle of the software we were evaluating.

The general lack of background research on institutional repositories had a significant impact on our choice of methodology. It is particularly important to note that we do not know who the end users of institutional repositories are; some proponents posit that the public will access repositories to read the research they pay for through public funding [12], while others focus on academic research uses or promotional uses [13]. Not only do we not know who our end users are, we do not know what their activities related to a repository are likely to be. We do not know whether they will find repositories using search engines or via institutional web sites; we do not know whether information seekers will come looking for a single paper or more general information; and we do not know what level of information seeking skills they will have. Knowing so little about information seekers and their specific tasks makes it very difficult to design representative tasks or select representative sample populations for user studies, and it also means that there are not yet standard heuristics that can be applied to repository interfaces.

Institutional context also affected our choice of method. We had a single usability analyst who has over five years' experience in digital library usability. This particular background means that the analyst is what Nielsen terms a 'double expert' [14] (expert in the software domain, and in usability) for institutional repositories, and that rather than finding fewer than 50 percent of usability problems (as could usually be expected of a lone evaluator), the evaluator in this study might be expected to find the majority of the problems in an institutional repository interface. Working with the analyst was a usability-minded developer, who made changes to the interfaces as they were suggested to avoid drawing out the development cycle. This meant that the user interface changed frequently and was not stable for user testing.

The final factor influencing our choice of usability evaluation method was the software development process. Swinburne is a test site for VTLS VITAL, with the added problem that when we began testing, the software was still under development and thus somewhat unstable. Not only were there stability problems with the software, but the development focus had until that point been on functionality, and thus there were a significant number of obvious usability problems. This combination of factors meant that conducting user testing was likely to be unproductive.

2.2 Methodology

The primary methodology chosen for this evaluation was heuristic analysis [15]. This method was chosen because it allows for evaluation without users, and also because it can be done concurrently with improvements being made. It was determined that each interface presented to end users by the software should be examined both as an individual entity, and in the context of the software as a whole. However, this method needed supplementation because institutional repositories are a new technology and consistency in particular is a problematic heuristic in these circumstances.

Because we do not know who our users are, and because the usability of other repository software is untested, instead of comparing VITAL to other repository systems, we compared it to journal database interfaces (used for similar tasks) and finding interfaces used by the general public, including Google™, eBay™ and Amazon™. This allowed us to effectively determine best practice in finding interfaces to which we could compare VITAL.

However, we did not rely solely on other interfaces to inform our evaluation. We also called on significant research in the information-seeking field to understand what users of information seeking systems in general need from a system for it to be usable. This background research is reviewed briefly in Section 3.

For some specific aspects of institutional repository usability, heuristics are not enough—notably the language needed to describe concepts that do not exist in other systems. For these aspects we relied on a section of a more general survey administered to library users to explore their understanding of library jargon.

So while heuristic analysis was our primary methodology, we relied heavily on background research and interfaces from other domains to inform this analysis, and on a user survey to inform aspects of the interface that needed empirical input.

3   Background Research

Because the research on institutional repository usability is so limited, we relied on research about information seeking behavior (described in Section 3.1) and on studies of other information systems (reviewed in Section 3.2) to inform our analysis. We describe the implications of this research for our analysis in Section 3.3.

3.1   Information Seeking Behavior

There are two models of information seeking behavior that are particularly relevant in an online context [16, 17]. While these models differ in structure, they are largely similar in practice, both involving a recursive cycle of searching and browsing. This understanding of information seeking behavior is borne out by studies of information seeking in on- and offline environments [10, 18, 19], and shows that information seeking interfaces ideally support interleaved searching and browsing.

3.2 Previous Studies of Information Systems

There is significant information available about how information seekers use a wide range of online information systems. This information can and should be used to inform usable design for repository systems.

Large usage studies of journal databases have shown that individual users visit infrequently, search within narrow subject areas, and view small numbers of articles [19-21]. Collectively, however, users of journal databases access a wide range of materials and download more articles than they cite [22, 23].

Studies of research-focused digital libraries confirm that information seekers download only a small number of articles [9, 24, 25]. These same studies demonstrate that—contrary to what we might expect in research focused digital libraries—researchers perform simple searches with only a small number of terms. They also do not use advanced search options or read help materials, nor do they read beyond the first page of search results.

The library terminology used in library interfaces is a significant hindrance to their usability [26], and repositories share terminology with other library interfaces. Terms shown to be problematic by the research [27, 28] and found in institutional repository interfaces include 'citation', 'special collection', 'bibliographic information', and 'proceedings'.

3.3 Implications for Repository Usability

The background research demonstrates that institutional repositories *must* be usable if they are to be used by information seekers; they are infrequent users, they do not read help files, and they quickly move on if they cannot find what they are seeking. To best promote successful information seeking, it is advisable to allow users to interleave searching and browsing, and to provide search functionality that provides relevant results without using advanced search options.

4   Example Usability Improvements to Swinburne Research Bank

The VTLS VITAL interface is considerably customizable, though there are some restrictions imposed by the underlying software design. Within the capabilities of the software, usability improvements were made throughout Swinburne Research Bank based on the results of a comprehensive heuristic analysis. The examples presented here demonstrate the interplay of basic heuristic analysis, domain specific knowledge, and an empirical survey in improving the Swinburne Research Bank interface. In Section 4.1 we will discuss changes to the navigational structure of the interface, which were based primarily on heuristic analysis; in Section 4.2 we will examine changes made to the search interface, which rely more on domain specific knowledge, and in Section 4.3 we analyze changes made to the interface based on empirical survey results.

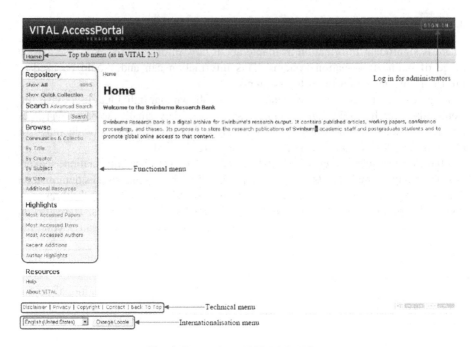

Fig. 1. The unaltered VITAL interface

4.1 Interface Navigation

The basic navigation structures of VITAL in its vanilla state were neither consistent with other search interfaces, nor appropriate given what we know about users (see Fig. 1). There were a number of menus spread over the interface, meaning that to adequately navigate the interface users had to remember where a number of different functions were located, creating large cognitive load. The main menu was at the left, taking up valuable screen real-estate, and forcing users to scroll considerably when viewing large amounts of content (which in a repository, or indeed any search interface, is not uncommon). While this left-hand location is consistent with DSpace (one of the other main repository software packages), it is inconsistent with other search interfaces, including EBSCOHost™, eBay™ and Google™. Other search interfaces usually use this space for second-level menu items such as browse menus and search facets or disambiguation links, causing flow-on consistency problems in VITAL, which offers both of these functionalities.

We made considerable changes to the basic navigation of the site (see Fig. 2) to make it consistent with other search interfaces, to reduce the cognitive load on users when looking for menu items, and to reduce scrolling. We also removed a significant number of items from the menu (either because they were rendered extraneous by institutional policy, such as the privacy link, or because they were not useful, for example the internationalization menu). These changes were primarily based on general user interface heuristics.

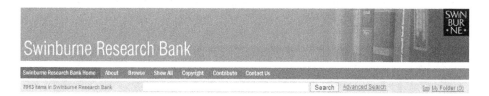

Fig. 2. The new navigational structure. All navigational elements to be used by information seekers are in one place, and extraneous items such as the internationalization menu and the technical menu have been removed.

4.2 The Search Interface

In VITAL's original state, the search box is approximately one third of the way down the page, to the left, and the advanced search link is next to the word 'Search'. This is a non-standard search box location and a very non-standard location for the advanced search link. Standard web guidelines recommend putting the search box in the top right or top left [29]; however in web interfaces designed for searching and likely to be used by academics or the general public (including Amazon™ EBSCOHost™, and Google™), there is a large central search box at the top of the page. The prominence of these search boxes provides users with a strong indication of how to interact with these systems, and afford entering multiple search terms, making it more likely that users' results will be relevant. Given that an institutional repository interface is primarily a finding interface, we have opted for consistency with other finding interfaces rather than general web guidelines. Again, consistent with the majority of finding interfaces, we have included a link to advanced search to the right of the search button (see Fig. 2).

VITAL presents search results in a non-standard way—they are presented in a table that can be re-sorted by clicking one of the table headings; the default sort is by date. While normally this inconsistent presentation would be a significant hindrance to good usability, in this case the ability to re-sort search results at will affords the interleaving of search and browsing that we identified as important in Section 3.2. In the original interface, however, there were no visual clues that the table headings were clickable (see Fig. 3), and minimal cues as to which element the results were sorted by (the gray in the background of the date column is slightly darker than the rest of the table). Ideally we would have changed the default result order to relevance ranking, because users rarely look beyond the first page of search results [9], however this was not feasible within the restrictions on customization. However we were able to provide stronger visual cues about the interactivity and sort order of search results by using a visual cue that is familiar to email users—the small triangle to the right of a column heading (see Fig. 4). Column headings were also underlined to look more like links, and therefore clickable.

Showing results 181 - 195 of 8597.

First Previous | 9 10 11 12 13 14 15 16 17 18 19 20 21 22 23 | Next Last

Title	Creator	Date	Full Text
Random Walk Smooth Transition Autoregressive Models	Anderson, Heather M.; Low, Chin Nam	2004	
Random Walk Smooth Transition Autoregressive Models	Anderson, Heather M.; Low, Chin Nam	2004	

Fig. 3. Original search result display

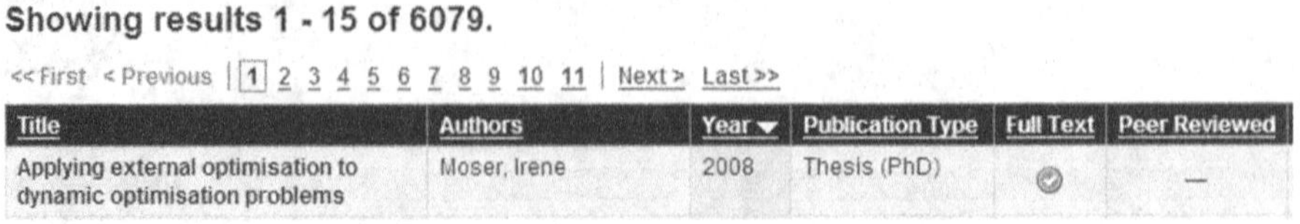

Showing results 1 - 15 of 6079.

<<First <Previous |1| 2 3 4 5 6 7 8 9 10 11 | Next> Last>>

Title	Authors	Year ▼	Publication Type	Full Text	Peer Reviewed
Applying external optimisation to dynamic optimisation problems	Moser, Irene	2008	Thesis (PhD)	⊘	—

Fig. 4. Improved result display, including arrow to indicate sort and link markings

VITAL provides search disambiguation, or 'facets', which users have been shown to like and use [30]. In the unaltered interface, however, these facets appeared underneath search results, and unlabeled., and when a search facet was clicked it moved to the left hand menu. There were a number of usability problems with the original facets: they were unlikely to be seen if users just glanced at results; they were in a location inconsistent with other search interfaces (for example EBSCOHost™ and eBay™); it was not clear what they were for; and the feedback when they were clicked was poor. Despite left menus increasing scrolling, we moved the facets to the left hand side of the search results (to be consistent with other search interfaces). We labeled them 'Limit search by', and improved the feedback by having selected facets appear in a new menu immediately above the ones that were not selected. These changes make the search facets easier to understand and more likely to be used.

The changes we made to the search interface relied not just on good heuristics, but also on knowing the information seeking domain and the research surrounding it.

4.3 Empirical Changes

For some aspects of the VITAL interface, notably terminology aspects, we could not rely on background research or good heuristics. As part of a larger survey of the university population about library jargon, we asked users about a number of aspects of Swinburne Research Bank that were either repository- or institution-specific.

One area that is particularly relevant to repositories is the display of metadata for each item; repositories traditionally use Dublin Core metadata, but we were not sure the element names VITAL was presenting to our users (Creator, Resource Type, Date, Source) would be easily comprehended. Based on the results of this survey, we changed the display of the first three those elements to 'Author', 'Publication type' and 'Publication year' respectively. We discovered that 'Source' was well understood (contrary to our expectations) and left it as it was.

We were also concerned about the display of subject metadata; as an Australian technological university, we were displaying RFCD (Research Fields Courses and Disciplines) codes, and plain text subject keywords. We were not sure what subject metadata users would find useful, nor how they would like it displayed. Based on the results of the survey, we decided to present subject keywords and RFCD codes separately, and include the numeric codes as part of the RFCD code display.

These terminology areas are relatively new and specific to institutional repositories—even databases typically do not mix different types of publications, and they have formal and rigid subject hierarchies. Because this is such a new field, we used empirical data for these specific questions that could not be answered by looking to the research, but which could be answered without exposing users to the VITAL interface during redevelopment.

5 Conclusions

Performing a heuristic analysis of web software when little is known about the likely users of the software or the tasks they will perform presents special challenges. In an industrial environment, where there is only a single usability analyst and pressure to make changes in a timely manner is particularly challenging.

In this paper we presented a case study of just such a heuristic analysis, and demonstrated how background research, empirical surveys, and comparison to similar systems can be used together judiciously to make significant usability improvements even to a novel type of interface. Despite the obvious obstacles to our analysis, we consider the method to have been successful in determining and implementing a number of usability improvements, demonstrating this approach as a useful first pass at making usability improvements in an industrial context.

6 Future Work

While significant improvements have been made to the Swinburne Research Bank interface, there is still scope for further work. If we understood who was likely to use institutional repositories, and how they were likely to use them, a more fine-grained heuristic analysis could yield further improvements. More importantly, though, this understanding could facilitate user testing with representative users and tasks, both to test the usability improvements already made, and to suggest further improvements.

The first stage in future work should be to conduct further study—including examining usage log studies—to determine what actually happens when users visit institutional repositories. This understanding would in turn facilitate empirical user testing to gain further insight into institutional repository usability.

Acknowledgements

Work on this paper was funded by the ARROW Project (Australian Research Repositories Online to the World), www.arrow.edu.au. The ARROW Project was funded by the Australian Commonwealth Department of Education, Science and Training, under the Research Information Infrastructure Framework for Australian Higher Education.

References

1. Harnad, S., Brody, T., Vallières, F., Carr, L., Hitchcock, S., Gingras, Y., Oppenheim, C., Stamerjohanns, H., Hilf, E.R.: The Access/Impact Problem and the Green and Gold Roads to Open Access. Serials Review 30, 310–314 (2004)
2. Henty, M.: Ten Major Issues in Providing a Repository Service in Australian Universities. D-Lib. 13 (2007)
3. Harnad, S.: Open Access To Peer Reviewed Research Through Author/Institution Self-Archiving: Maximizing Research Impact by Maximizing Online Access. Journal of Postgraduate Medicine 49, 337–342 (2003)

4. Fried Foster, N., Gibbons, S.: Understanding Faculty to Improve Content Recruitment for Institutional Repositories. D-Lib. 11 (2005)
5. Kim, J.: Motivating and Impeding Factors Affecting Faculty Contribution to Institutional Repositories. In: Joint Conference on Digital Libraries, Chapel Hill, NC, USA. ACM Press, New York (2006)
6. Kim, J.: Finding Documents in a Digital Institutional Repository: DSpace and ePrints. In: 68th Annual Meeting of the American Society for Information Science and Technology, Charlotte, North Carolina, American Society for Information Science and Technolgy (2006)
7. Bell, S.J.: The Infodiet: How Libraries Can Offer an Appetizing Alternative to Google. Chronicle of Higher Education 50, 15 (2004)
8. De Rosa, C., Cantrell, J., Cellentani, D., Hawk, J., Jenkins, L., Wilson, A.: Perceptions of Libraries and Information Resources. OCLC, Dublin, Ohio, USA (2005)
9. Jones, S., Cunningham, S.J., McNab, R.: An Analysis of Usage of a Digital Library. In: European Conference on Digital Libraries, Heraklion, Crete, pp. 261–277. Springer, Heidelberg (1998)
10. Nordlie, R.: User Revealment — A Comparison of Initial Queries and Ensuing Question Development in Online Searching and in Human Reference Interactions. In: 22nd Annual ACM Conference on Research and Development in Information Retrieval, pp. 11–18. ACM Press, Berkeley (1999)
11. Agosto, D.E.: Bounded Rationality and Satisficing in Young People's Web Based Decision Making. Journal of the American Society for Information Science and Technology 53, 16–27 (2002)
12. Lynch, C.A.: Institutional Repositories: Essential Infrastructure for the Scholarship in the Digital Age. Bimonthly Reports. Association of Research Libraries (2003)
13. Crow, R.: The Case for Institutional Repositories: A SPARC Position Paper. The Scholarly Publishing and Academic Resources Coalition (2002)
14. Nielsen, J.: Finding Usability Problems Through Heuristic Evaluation. In: Proceedings of the SIGCHI conference on Human factors in computing systems, Monterey, California, United States. ACM, New York (1992)
15. Nielsen, J., Molich, R.: Heuristic Evaluation of User Interfaces. In: Proceedings of the SIGCHI conference on Human factors in computing systems: Empowering people. ACM, Washington (1990)
16. Kuhlthau, C.C.: Inside the Search Process: Information Seeking from the User's Perspective. Journal of the American Society for Information Science and Technology 42, 361–371 (1999)
17. Marchionini, G.: Information Seeking in Electronic Environments, vol. 9. Cambridge University Press, Cambridge (1995)
18. Crabtree, A., Twidale, M.B., O'Brien, J., Nichols, D.M.: Talking in the Library: Implications for the Design of Digital Libraries. In: Second International ACM Conference on Digital Libraries, pp. 221–228. ACM Press, Philadelphia (1997)
19. Nicholas, D., Huntington, P., Jamali, H.R., Tenopir, C.: Finding Information in (Very Large) Digital Libraries: A Deep Log Approach to Determining Differences in Use According to Method of Access. Journal of Academic Librarianship 32, 119–126 (2006)
20. Nicholas, D., Huntington, P., Jamali, H.R., Watkinson, A.: The Information Seeking Behaviour of the Users of Digital Scholarly Journals. Information Processing and Management 42, 1345–1365 (2006)
21. Nicholas, D., Huntington, P., Watkinson, D.: Scholarly Journal Usage: The Results of Deep Log Analysis. Journal of Documentation 61, 248–280 (2005)

22. Huntington, P., Nicholas, D., Jamali, H.R., Tenopir, C.: Article Decay in The Digital Environment: An Analysis of Usage of OhioLINK by Date of Publication, Employing Deep Log Methods. Journal of the American Society for Information Science and Technology 57, 1840–1851 (2006)

23. Nicholas, D., Huntington, P.: Electronic Journals: Are They Really Used? Interlending and Document Supply 34, 48–50 (2006)

24. Mahoui, M., Cunningham, S.J.: Search Behavior in a Research Oriented Digital Library. In: European Conference on Digital Libraries, pp. 13–24. Springer, Germany (2001)

25. Mahoui, M., Cunningham, S.J.: A Comparative Transaction Log Analysis of Two Computing Collections. In: Baker, J.B.a.T. (ed.) European Conference on Digital Libraries, Lisbon, Portugal, pp. 418–423. Springer, Heidelberg (2000)

26. Augustine, S., Greene, C.: Discovering How Students Search a Library Web Site: A Usability Case Study. College and Research Libraries 63, 354–365 (2002)

27. Chaudhry, A.S., Choo, M.: Understanding of Library Jargon in the Information Seeking Process. Journal of Information Science 27, 343–349 (2001)

28. Hutcherson, N.B.: Library Jargon: Student Recognition of Terms and Concepts Commonly Used by Librarians in the Classroom. College and Research Libraries 65, 349–354 (2004)

29. Johnson, J.: GUI Bloopers 2.0: Common User Interface Design Don'ts and Dos. Morgan Kaufman, San Francisco (2007)

30. Ka-Ping, Y., Kirsten, S., Kevin, L., Marti, H.: Faceted metadata for image search and browsing. In: Proceedings of the SIGCHI conference on Human factors in computing systems, Ft. Lauderdale, Florida, USA. ACM, New York (2003)

Measuring Web Accessibility by Estimating Severity of Barriers

Giorgio Brajnik

Dip. di Matematica e Informatica
Università di Udine
Italy
www.dimi.uniud.it/giorgio

Abstract. The paper addresses the issue of measuring web accessibility in such a way that differences in measurements reflect differences in the effectiveness experienced by disabled users. The paper presents the steps upon which a measuring methodology called MAMBO is based, and the data that are needed to compute the indexes, in addition to its conceptual rationale. An experimentation of MAMBO is then described, based on analysis of 14 accessibility reports; results are shown and discussed, including the effects that different severity judgments may have on the metric, how to estimate confidence intervals on the values, and how the metric can be used to estimate accessibility with respect to specific user groups.

1 Introduction

The importance of measuring web accessibility is increasing; many different activities are demanding it. For example, measuring takes place when quality assurance practitioners are monitoring accessibility of a website to ensure that it does not decrease as new content gets published. Similarly, measuring occurs when developing a new user interface of the website and comparing it to previous versions of the same website, or to websites of competitors (competitive analysis). Once accessibility defects are found, in order to set priorities, developers need to know which ones are more important in terms of negative impact on users experience: a measure of accessibility is again needed. Accessibility levels are also needed when end users want to know how accessible a website is before using it. This is the case, for example, when a search engine lists search results that are ranked also by accessibility level [2].

Many existing measurement processes are based on the number of violations of established requirements (for example, WCAG 1.0 checkpoints). While the conformance level of the website (*i.e.* the degree to which a website satisfies the requirements defined by a standard) is an important measure when formal regulations are in place, this is by no means the only way to determine the accessibility level. One limit of methods based on conformance is that it is difficult to relate the accessibility level to the actual hindrances that the website may raise against given user categories, such as blind users of screen readers,

S. Hartmann et al. (Eds.): WISE 2008, LNCS 5176, pp. 112–121, 2008.

low-vision users of screen magnifiers, motor-disabled users of a normal keyboard and/or mouse, deaf users, cognitively disabled users (with reading and learning disabilities and/or attention deficit disorders). Yet, this is what is often meant by *accessibility*: "a web site is accessible if people with some impairment can use it with the same effectiveness, security and safety as non-disabled people" (definition derived from [10]). If accessibility levels were determined with this definition in mind, and by applying a *valid* measurement process[1], then users could easily determine how much accessible the website is for them; developers and QA practitioners could estimate which user categories could be best or worst supported by the website; they could determine which parts of the website do a better (or worse) job in supporting these users; they could compare different versions of the website to determine how accessibility is evolving along the development; and they could set fixing priorities.

Several accessibility metrics have been discussed in the literature [1, 3, 11, 12, 14]. In many cases the measurement process is based on automated testing tools, capable of systematic application of an array of tests covering some or all of the requirements of a standard. The advantages behind such a solution is that tools are systematic scanners of web pages, efficient processors and reliable evaluators (in the sense that they produce repeatable results). However, tools are plagued with the problems of generating issues that are not accessibility problems (false positives), of missing certain true accessibility problems (false negatives), and of being incapable of estimating the severity of a requirement violation. Since the tools do conformance testing, they yield a measure of accessibility that is a function of the number of passed and failed requirements/tests. Often this function is the *failure rate* FR, defined as the number of violations of any checkpoint divided by the maximum number of violations of any of those checkpoints that can take place (*i.e.* by the number of *possible* violations). For example, two pages that include 10 and 20 images respectively, one with 2 properly defined "alt-text", the other with 8, have FR = 0.8 and FR = 0.6 respectively. Even though the second page has a larger number of violations, hence a larger number of potential obstacles to users, it has a smaller FR. Therefore, in addition to wrong estimates due to false positives and negatives, the values produced by accessibility metrics based on automatic tools cannot be directly be related to accessibility as defined above.

For these reasons, we defined and experimented SAMBA [5], a method for measuring accessibility by using the output of testing tools *coupled with* focused opinions of experienced human evaluators, so that correct estimates of tool errors can be assessed, and appropriate estimations of severities of barriers are used. When adopting SAMBA, an accessibility testing tool is used to automatically test many web pages; this generates a large number of checkpoint violations, that are automatically mapped to potential accessibility barriers and then sampled

---

[1] *Validity* is "the extent to which the problems detected during an evaluation are also those that show up during real-world use of the system" whereas *reliability*, often called *reproducibility*, is "the extent to which independent evaluations produce the same result" [8].

randomly. The sample of potential barriers is submitted to a panel of judges that assign them a severity level, including 0 ("not-a-problem").

This paper describes a manual method for measuring barriers of accessibility (MAMBO) that derives from SAMBA and that can be used when manually evaluating accessibility of a website. The accessibility indexes defined by MAMBO are standardized (and therefore can be used to compare accessibility levels of different websites and/or obtained by different evaluators). In addition, MAMBO offers more than one accessibility index, which can be used to measure accessibility with respect to different user groups, and to estimate the uncertainty due to having analyzed only a fraction of the available web pages. To adopt MAMBO evaluators follow an assessment method called *barrier walkthrough* [4]; computing the accessibility indexes requires little additional effort.

2   Barrier Walkthrough

The barrier walkthrough (BW) method [4, 6] is an accessibility inspection technique. An evaluator has to consider a number of predefined barriers which are interpretations and extensions of well known accessibility principles; they are linked to user characteristics, user activities, and situation patterns so that appropriate conclusions about user effectiveness, productivity, satisfaction and safety can be drawn, and appropriate severity scores can be consequently derived. The method is rooted on heuristics walkthrough [9] which considers the context of use of the website. For BW, context comprises certain user categories (like blind persons), usage scenarios (like using a given screen reader), and user goals (corresponding to *use cases*, like submitting an IRS form).

An *accessibility barrier* is any condition that makes it difficult for people to achieve a goal when using the web site through specified assistive technology (see figure 1 for an example). A barrier is a failure mode of the web site, described in terms of (i) the user category involved, (ii) the type of assistive technology being used, (iii) the goal that is being hindered, (iv) the features of the pages that raise the barrier, and (v) further effects of the barrier.

Notice that several barriers can depend on the same cause: *e.g.* for a missing *skip-links* link (defect) a barrier for a blind user of a screen reader is that s/he cannot get quickly to the relevant content of the page; the barrier for a keyboard user is that s/he cannot move the focus directly to the relevant controls in the page; the barrier for a low vision person is that s/he cannot move directly the field of vision on the relevant content.

Severity of a barrier depends on the context of the analysis (type of user, usage scenario, user goal). The BW method prescribes that severity is graded on a 1–2–3 scale (minor, major, critical), and is a function of *impact* (the degree to which the user goal cannot be achieved within the considered context) and *frequency* (the number of times the barrier shows up while a user is trying to achieve that goal). Therefore the same type of barrier may be rated with different severities in different contexts; for example, the missing *skip-links* link may turn out to be a nuisance for a blind user reading a page that has few preliminary stuff, while the same defect may show a higher severity within a page that does a server refresh whenever the user interacts with links or select boxes.

Potential barriers to be considered are derived by interpretation of relevant guidelines and principles [7, 13]. A complete list can be found in [6].

barrier	users cannot perceive nor understand the information conveyed by an information rich image (*e.g.* a diagram, a histogram)
defect	an image that does not have accompanying text (as an ALT attribute, content of the OBJECT tag, as running text close to the picture or as a linked separate page)
users affected	blind users of screen readers, users of small devices
consequences	users try to look around for more explanations, spending time and effort; effectiveness, productivity, satisfaction are severely affected

Fig. 1. Example of barrier

3 MAMBO

MAMBO (MAnually Measuring Barriers Of accessibility) is an accessibility metric not based on conformance. It can be computed directly by scanning a barrier walkthrough report, by highlighting reported barriers, and by considering their severity and the kind of user groups to which they refer (blind persons, motor disabled ones, etc.).

The basic computation of the accessibility index (AI) is similar to SAMBA [5]. In particular (Table 1 shows some example):

1. By tabulating the number of barriers split by user types against each severity value, we obtain the *severity matrix*; each element of the matrix M gives the proportion of sampled barriers associated with disability d and severity s.
2. The *confidence intervals severity matrix* $\mathcal{M}$ can then be generated, by computing the 95% confidence interval around each proportion $M_{d,s}$.
3. The *barrier density* of a web site needs to be computed. It is defined as $F = k\frac{\text{number of barriers}}{\text{num. of bytes}}$, which can be interpreted as the probability that k bytes of HTML code of the site causes a barrier; if M is the severity matrix, then $F \cdot M_{d,s}$ is the probability that k bytes of code causes a barrier for disability d with severity s; the scale factor k is used to tune the values produced by MAMBO.
4. If we combine the density factor F with the confidence interval severity matrix $\mathcal{M}$ we obtain $\mathcal{F}$; after using appropriate weights to balance different severity levels, we get the *Weighted Accessibility Index* (AI_w). Since it is based on confidence intervals, it is itself an interval $(\underline{AI_w}, \overline{AI_w})$, defined as follows:

$$\text{let } \underline{H_d} = \frac{\underline{f}_{d,1}}{w_1} + \frac{\underline{f}_{d,2}}{w_2} + \underline{f}_{d,3}$$

$$\text{and } \overline{H_d} = \frac{\overline{f}_{d,1}}{w_1} + \frac{\overline{f}_{d,2}}{w_2} + \overline{f}_{d,3}$$

$$\text{then } \underline{AI_w} = \prod_d \left(1 - F \cdot min\left\{1, \overline{H_d}\right\}\right)^2$$

$$\text{and } \overline{AI_w} = \prod_d \left(1 - F \cdot min\left\{1, \underline{H_d}\right\}\right)^2$$

where $\mathcal{F}[d, s] = f_{d,s}$, and $\frac{1}{w_s}$ is the weight to be given to minor and major barriers (*i.e.* $s = 1, 2$). Each term in the product defining AI_w can be interpreted as the probability that no barriers for disability d are raised, and the resulting value is related to the probability that there are no barriers at all. Squaring the terms further amplify the contribution of each disability.

For example, the severity matrix illustrated in Table 1 shows that 35 barriers for blind users were found; 14% were minor ones, 23% major and 63% were found to be critical. The table shows also the confidence intervals around these proportions; for example, it is safe to assume that critical barriers for blind users range between 45% to 78%.

Table 1. (Left) Severity matrix obtained from a barrier walkthrough report. Columns 1 to 3 show the proportion of barriers that were given severity 1 to 3 (*minor* to *critical*); the last column gives the total number of barriers. (Right) Confidence interval matrix from the same report ($\alpha = 0.05$).

category	Severity 1	2	3	tot	category	Severity 1	2	3
cb (color blind)	0.00	0.00	1.00	**2**	cb	0.00 0.80	0.00 0.80	0.20 1.00
md (motor disab.)	0.11	0.47	0.42	**19**	md	0.02 0.35	0.25 0.71	0.21 0.66
lv (low vision)	0.00	0.00	0.00	**0**	lv	-	-	-
nh (no hearing)	0.00	0.00	0.00	**0**	nh	-	-	-
nv (no vision)	0.14	0.23	0.63	**35**	nv	0.05 0.31	0.11 0.41	0.45 0.78
cd (cogn. disab.)	0.36	0.32	0.32	**25**	cd	0.10 0.46	0.28 0.68	0.10 0.46
js (no javascript)	0.00	0.33	0.67	**9**	js	0.00 0.37	0.09 0.69	0.31 0.91

For the same report the barrier density factor is 0.039 (barriers/k bytes of code, with $k = 20$); using weights $1/9, 1/3$ (one critical barrier weighs as much as 9 minor and 3 major ones), if we restrict to the *no-vision* category, we obtain $AI = 0.94$, and an interval of $(0.92, 0.96)$. After combining all the disability types we get $AI = 0.88$ and an interval of $(0.68, 0.75)$.

4   Practical Examples and Discussion

A practical analysis of MAMBO was carried out by analyzing 14 barrier walkthrough reports produced by students of my course (user centered web design).

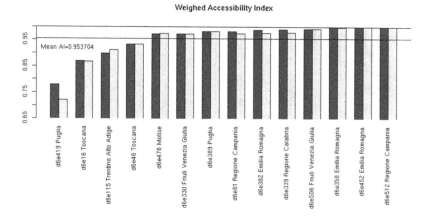

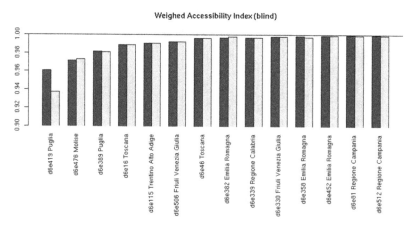

Fig. 2. Web Accessibility Index for the 14 reports (some refer to the same web site, Italian local government web sites; for example, "d6e382 Emilia Romagna" and "d6e358 Emilia Romagna" are two different reports about the same web site). (*Top*) The dark area is the index derived from the judged severity; the other one derives from the original severity scores. The horizontal line gives the mean index (0.9537). (*Bottom*) Same indexes, but with data restricted to barriers for blind users.

Students were exposed to web accessibility, conformance testing and barrier walkthrough for about 15 lecture hours, after which they were asked to analyze given web sites and write corresponding reports[2].

These BW reports were analyzed by a judge, who was asked to validate the severity judgments made by the authors of reports. The judge had to give her own severity level to each of the barriers mentioned in the report. In this way we

---

[2] The entire set of Italian reports is available at www.dimi.uniud.it/giorgio/dida/psw/galleria/galleria.html

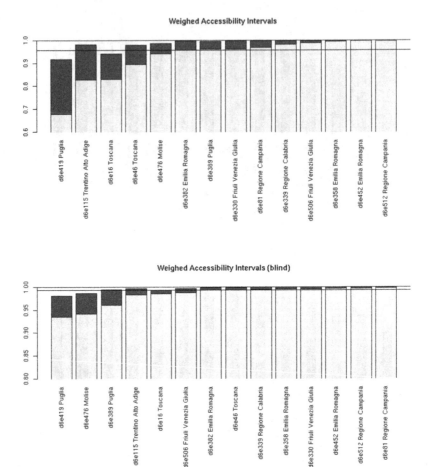

Fig. 3. (*Top*) The intervals for AI on the same reports (the dark area represent lower and upper bounds of AI), based on judge's severity. (*Bottom*) The intervals for AI on the same reports restricted to barriers for blind users.

can estimate what is the effect of *false positives* on the metric. The values given below were computed using the severity that the judge assigned to problems, including 0 for what was deemed to be a "not-a-problem".

Figure 2 (top) shows that AI spans a relatively small range (0.72 to 0.99); this is due to the magnitude of the density factor: the smaller it is, the wider is the range spanned by AI. Therefore, by using an appropriate scale factor k, results can be tuned to the desired level of resolution. More importantly, Figure 2 shows that different judgments of barriers severity have limited effects on the overall AI value, even though over 327 analyzed barriers, there were 59 disagreements in severity (18%).

When restricting to barriers relevant for a given user group, for example blind users (bottom part of Figure 2), the AI range narrows (since fewer barrier types are considered, and fewer disability-related contributions to AI are included). Also in this case the effect of misjudged severity is marginal.

Figure 3 (top) shows the intervals around AI, computed on the basis of the 5% confidence intervals of the severity matrix. When two AI intervals overlap then no comparison can be made on the corresponding websites (or reports), since the level of uncertainty determined by the number of found barriers and their splitting into different severities is too high. However, when two intervals do not overlap a comparison between web sites/reports can be stated with relatively high certainty. For example, only two sites are less accessible than the sixth one (d6e382 Emilia Romagna): those whose upper bound is below the horizontal line. The certainty level of this statement is higher than 95%.

Obviously, when restricting to fewer disability types, the range narrows; this is shown at the bottom part of the Figure. But also in this case non-overlapping intervals can be used to compare web sites. For example, two websites are less accessible than the seventh one (d6e382 Emilia Romagna).

Although reducing the precision of the metric, intervals are useful to represent accessibility indexes when we know before hand that only a fraction of the website pages were analyzed. Intervals, in such cases, lead to comparison statements that have a measurable level of certainty.

5 Conclusions

MAMBO is a metric that can be used with data gathered from barrier walkthrough accessibility reports, containing estimates of the severity of accessibility barriers. Using these estimates, the probabilistically-based schema used by MAMBO leads to sound accessibility indexes.

MAMBO is flexible: it can be used for generically comparing websites or accessibility reports; for numerically estimating the effects of judgment errors; and for estimating the uncertainty levels due to an accessibility investigation that was limited to few pages. Provided that appropriate severity judgments are applied also on conformance reviews (like those based on WCAG 2.0), then MAMBO can be used on those reports as well.

The level of experience in accessibility of web technologies, in accessibility evaluations, and in assistive technologies obviously affect the outcome of MAMBO. Although appearing robust with respect to judging mistakes, by providing incorrect ratings of severities any results can be produced, and MAMBO has no intrinsic correction mechanism. Incorrect rating could reflect more false negatives, more false positives, different distributions of these among user categories, or unbalanced judgment of severities for the true positives. However, the study we reported here was performed on reports written by novice evaluators where a substantial number of judging mistakes were made, and nevertheless the confidence intervals produced through MAMBO were shown to be relatively small.

But this study was limited to false positives; a limit of MAMBO is its inability to cope with false negatives, *i.e.* accessibility barriers that are missed by evaluators. We plan to investigate if appropriate merging of semi-automatic and manual evaluation techniques can provide some reliable estimate of false negatives. Another future research opening is to set up a web accessibility observatory based on MAMBO and SAMBA and to determine how MAMBO correlates with SAMBA and with failure-rate based metrics, such as WAQM.

Acknowledgements

I would like to thank Chiara Cecotti for her help in judging students' reports, and the anonymous reviewers for their helpful suggestions.

References

[1] Arrue, M., Vigo, M., Abascal, J.: Quantitative metrics for web accessibility evaluation. In: Proc. of the ICWE 2005 Workshop on Web Metrics and Measurement (2005)

[2] Arrue, M., Vigo, M., Abascal, J.: Web accessibility awareness in search engines results. Int. Journal on Universal Access in the Information Society (2008); In press

[3] Bailey, J., Burd, E.: Tree-map visualization for web accessibility. In: COMPSAC 2005: Proceedings of the 29th Annual International Computer Software and Applications Conference (COMPSAC 2005), vol. 1, pp. 275–280. IEEE Computer Society, Washington (2005)

[4] Brajnik, G.: Web Accessibility Testing: When the Method is the Culprit. In: Miesenberger, K., Klaus, J., Zagler, W., Karshmer, A. (eds.) ICCHP 2006. LNCS, vol. 4061, pp. 156–163. Springer, Heidelberg (2006)

[5] Brajnik, G., Lomuscio, R.: SAMBA: a semi-automatic method for measuring barriers of accessibility. In: Trewin, S., Pontelli, E. (eds.) 9th Int. ACM SIGACCESS Conference on Computers and Accessibility, ASSETS, Tempe, AZ, October 2007. ACM Press, New York (2007)

[6] Brajnik, G.: Web accessibility testing with barriers walkthrough (March 2006) (Visited May 2008),www.dimi.uniud.it/giorgio/projects/bw

[7] DRC. Formal investigation report: web accessibility. Disability Rights Commission (March 2006) (Visited January 2006),
www.drc-gb.org/publicationsandreports/report.asp

[8] Gray, W.D., Salzman, M.C.: Damaged merchandise: a review of experiments that compare usability evaluation methods. Human–Computer Interaction 13(3), 203–261 (1998)

[9] Sears, A.: Heuristic walkthroughs: finding the problems without the noise. Int. Journal of Human-Computer Interaction 9(3), 213–234 (1997)

[10] Slatin, J., Rush, S.: Maximum Accessibility: Making Your Web Site More Usable for Everyone. Addison-Wesley, Reading (2003)

[11] Sullivan, T., Matson, R.: Barriers to use: usability and content accessibility on the web's most popular sites. In: Proc. of ACM Conference on Universal Usability, pp. 139–144 (2000)

[12] Velleman, E., Velasco, C.A., Snaprud, M., Burger, D.: D-WAB4 Unified Web Evaluation Methodology (UWEM 1.0). Technical report, WAB Cluster (2006)

[13] W3C/WAI. How people with disabilities use the web. World Wide Web Consortium — Web Accessibility Initiative (March 2004),
http://w3.org/WAI/EO/Drafts/PWD-Use-Web/20040302.html
[14] Zeng, X.: Evaluation of Enhancement of Web Content Accessibility for Persons with Disabilities. PhD thesis, University of Pittsburgh (2004)

Modelling Web-Based Systems Requirements Using WRM*

Fernando Molina[1], Jesús Pardillo[2], and Ambrosio Toval[1]

[1] Department of Informatics and Systems
University of Murcia
{fmolina,atoval}@um.es
[2] Department of Software and Computing Systems
University of Alicante
jesuspv@dlsi.ua.es

Abstract. In recent years, Web Information Systems (WIS) development projects have grown increasingly complex and critical for the
smooth running of the organizations. However, recent studies reveal that
a high percentage of web-based development projects miss the quality parameters required by stakeholders, most of the time due to an incorrect
requirements management. In this paper, we try to increase the weight
of requirements engineering activities in Web Engineering, and propose a
web engineering requirements metamodel extension that can be smoothly
integrated with existing web engineering proposals in order to reinforce
the first phases of systematic web development. Our proposal is accompanied by a tool that, being developed as an Eclipse plug-in, can also
be integrated with any existing web engineering methodology developed
under this general framework.

1   Introduction and Motivation

The development of Web Information Systems (WIS) has lived an exponential
growth in the last decade. Initially, these systems were used only as a means
to disseminate information. However, nowadays their complexity has increased,
they are present in numerous domains (electronic bank, healthcare, etc.) and
they have become critical systems for the business strategy of many organizations [1]. For these reasons, organizations have adapted their software development processes to deal with the idiosyncrasy of web applications [2] and the
research community is developing numerous methodologies within the scope of
a new discipline called Web Engineering (WE) [3], with the aim of helping in
the development of successful WIS.

However, as several surveys [4] reveal, the development of this kind of systems
is not exempt from errors, and the WIS finally developed do not always satisfy

* Partially supported by the Spanish projects DEDALO (TIN2006-15175-C05-03),
ESPIA (TIN2007-67078), DADS (PBC-05-012-2) and the network CALIPSO
(TIN2005-24055-E). Fernando Molina is funded under an FPI grant (Fundación
Séneca, Región de Murcia) and Jesús Pardillo under the FPU grant AP2006-00332.

S. Hartmann et al. (Eds.): WISE 2008, LNCS 5176, pp. 122–131, 2008.

the quality requirements demanded by their users. Specifically, [4] highlights that the top five problem areas of large-scale web application projects are (1) failure to meet business needs (84%), (2) project schedule delays (79%), (3) budget overrun (63%), (4) lack of required functionality (53%) and (5) poor quality of deliverables (52%). All these problems, far from being new, are quite similar to those encountered in traditional Software Engineering, where it has already been proven [5] that they are often a symptom of an inadequate management of the tasks related to the requirements discipline of the project.

This avowed influence of Requirements Engineering (RE) in the success of a project is even more paramount in web based development projects. In [1], a set of interviews with organizations that develop web-based systems was carried out and the majority of respondents (76%) indicated the importance of gathering the right requirements as a critical factor for the success of their web based development projects. Also other authors outline that the fact that web applications are more sensible to aspects such as ease of navigation, usability, accessibility and so on creates a necessity for them to follow a RE process even more detailed than the one followed during traditional software developments [2,6].

In spite of these considerations, the problems related to requirements management in web-based development projects do not seem to be solved neither in the organizations that carry out these projects nor in the methodologies proposed for WIS development. A survey carried out by [2] over 160 organizations that develop web-based software reveals that, with regard to requirements, the 60% of the organizations consider that one of the main problems in their projects is related to the clarity and stability of the requirements. On the other hand, Web Engineering methodologies are mostly focused on the design of the web applications, while RE activities are, in the best of cases, just tangentially tackled [6]. A recent effort [7] to homogenize the concepts managed in WE has only partially solved this situation because the metamodels developed are again focused on concepts too near of the WIS design - such as navigation structures or presentation features, to name a few- and therefore they are not suitable to cover the more abstract concepts managed by users or modellers.

On the other hand, WIS must pay special attention to some quality requirements such as usability, accessibility, etc. However, [8] remarks that their elicitation is usually implicitly understood for the stakeholders and it can usually lead to problems with their satisfaction on the delivered products.

In view of this situation, and since empirical data demonstrates that efforts invested in an adequate requirements management considerably reduce drawbacks in later phases of the development and improve the productivity and quality of the processes and software products [9], this paper proposes a reinforcement of WIS development activities related to requirements management. To achieve this goal, we propose a Web Requirements Metamodel (WRM) that can be easily integrated with existing WE proposals, together with an Eclipse tool that supports it. In this way, we aim at easing the integration of our proposal with existing web engineering methodologies. Our believe is that this integration can help to improve the requirements management in web based development projects. This

RE process enforcement may, in turn, help to reduce the number of failures detected in WIS development projects, with the final aim of increasing the quality of the WIS finally developed and the satisfaction of their users.

The rest of the paper is organized as follows. In Section 2, an overview of the related work and the main advantages of requirements metamodelling are provided. Section 3 describes WRM and the different concepts that make it up. Section 4 presents a brief study case that illustrate the instantiation of WRM. The automatic support developed to manage WRM is shown in Section 5. Finally, in Section 6, the conclusions are given and further research is outlined.

2   Requirements Metamodelling: Need, Advantages and Related Work

A requirements metamodel that defines in a formal way the concepts and relationships involved in the RE has been a long-sought aim by both researchers and organizations in Software Engineering, and now also in WE. The advantages that a metamodel offers are numerous [10]. On the one hand, the metamodel defines both the elements that participate in the requirements management process and their relationships in an unambiguous way. Moreover, the metamodel offers a formal basis on which tools for (1) the management of the metamodel elements and (2) the definition of transformation rules from requirements to other elements can be constructed [11]. Furthermore, most existing methodologies for WIS development have already been aligned with the MDA proposal [12,13,14]. Metamodelling is a key activity in MDA, and therefore the existence of a requirements metamodel that can be integrated with the general WE metamodel is a needed step to complement the methodologies with requirements management activities, thus emphasizing the role that requirements should pay in the WIS development process.

To our knowledge extent, this topic of requirements metamodelling in WE has only been tackled in [15,16]. With regard to [15], due to the design-oriented approach followed by most WE approaches, the concepts that appear in the metamodel are excessively near to WIS design (for instance, navigation nodes as well as search and navigation structures). Other kind of requirements related to the functionality of the WIS or even related to important attributes in web projects such as usability or accessibility can not be expressed. The main problem of this lack of high-level expressivity is that it is a generally avowed fact that design requirements are difficult to understand by stakeholders not directly related to design. Such stakeholders require a more abstract way to express their own requirements, that is, a way that is closer to the domain under which the application is being developed. The necessity of capturing high-level communication goals and user requirements is remarked in [16]. Their authors use a goal-oriented approach to model web requirements. The concept of goal was defined in the i* framework [17] and it models a high-level objective of one or more stakeholders. The concepts in the i* framework are useful to model user goals, although their generality suggests the necessity of tailoring them to specific domains. Following this trend, [16] uses i* as basis, but specializes it to design a new requirements metamodel that collects particular WE concepts. Although it gives a first step in web requirements metamodelling, [16] does not deal with other web requirements concepts such as the different techniques used by WIS development methodologies to refine requirement descriptions or the methods used for requirements validation. Moreover, it remarks the necessity of improving the automatic support and dealing with the concept of requirements reuse. These considerations, as well as the concept of goal defined in i*, have been included in the definition of WRM, which considers requirements validation methods and includes an automatic support to deal with the metamodel elements.

With regard to requirements metamodelling not specifically focused on web-based systems, two proposals have been specially influential in our approach. The first one is COMET [18], a requirements modelling method that includes a requirement metamodel. However, this metamodel does not pay attention to non-functional requirements. Moreover, it includes the UML use cases as the only requirements specification method and it does not cover any method for requirements validation. The second approach we would like to stress out is REMM [19], which presents a RE metamodel that includes the elements that usually appear in a requirements model. However, this generic metamodel does not consider some specific concepts related to web-based projects, such as specifications techniques or validations tools. Additionally, it suffers from a lack of full support for some non-functional requirements such as ease of navigation or accessibility. Despite this fact, some ideas in [19] (e.g. the concept of reuse, which will be explained later), are very useful and can be adapted to the WIS scope.

Summarizing, all the proposals presented so far include interesting ideas, some of which have been incorporated to our approach. However, the main disadvantage of them is that they do not deal with the specific needs of web-based projects and the specific requirements of this kind of projects. The next section further elaborates on these needs and how our Web Requirements Metamodel (WRM) answers them by reflecting both the concepts involved in a web-based project and the idiosyncrasy of its requirements.

3   Web Requirements Metamodel (WRM)

WRM is a requirements metamodel designed for the needs of web-based projects. For that, WRM syntethizes and simplifies the, from our point of view, most relevant concepts included in well-known RE proposals. Such simplification was needed to avoid the burden of work usually added by more exhaustive RE practices, on the premises that, in the WE community, baselines must be generated very quickly and therefore straightforward ways to gather requirements and connect them with validation methods are needed. Moreover, it stresses the importance of performing requirements management activities as a fist-order workflow (and not a tangential one, as it occurs nowadays) for web methodologies.

Figure 1 shows WRM. The key elements in WRM are requirements. Each requirement can be described using a set of attributes (hidden in the Figure 1) such as an *identifier*, the *textual description* of the requirements and so on. In order to avoid as much as possible the ambiguity inherent to natural language, the description of requirements can use well-defined *terms* included in a *glossary*. As Figure 1 shows, WRM divides requirements into *functional requirements* (what the system must do) and *non-functional requirements* (how the system must do it). This classification is useful because, while functional requirements usually rely on *test cases* to be validated, non-functional requirements are related to *quality scenarios*. More complex requirements classifications have been avoided, as the possibilities are countless, and greatly depend on the preferences of the designer. Requirements organization into tree-like structures (see e.g. quality models such as the ISO 9126 [20]) can be defined in a simple manner by means of the unary relationship *decomposedInto* defined over the requirement concept.

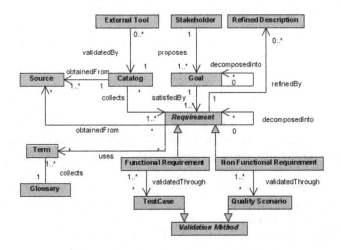

Fig. 1. WRM:Web Requirements Metamodel

Each requirement has a set of relationships with other elements. WRM includes the i* concept of *goal*, as a means to model high-level objectives of stakeholders. Each goal is related to the *stakeholder* that proposes it. The goal is satisfied through the fulfilment of a set of requirements. In the requirement element, the attribute description is used to describe a requirement. However, this description could be incomplete or ambiguous. Due to this fact, methodologies for WIS development need more precise techniques to describe requirements. These techniques include, but are not limited to, UML use cases (used for instance in UWE [13] or WebML[12]), scenarios (used in UWE [13]), specific notations (e.g. the notation NDT [21]) and methodology-dependent templates (e.g. those used in NDT [21]). Interested readers can find a more detailed description on how WIS methodologies model requirements in [6]. In order to support these techniques in a general manner, WRM includes the *refined description* element.

On the other hand, WRM introduces the element *catalog* (related to the concept of requirements reuse) to represent a set of related requirements. In a nutshell, a catalog puts together a set of requirements extracted from the same source, for example, a law, a guideline, etc. and can be reused in all the projects where these guidelines are applicable. This concept has been applied successfully in traditional requirements metamodels [22] and its adaptation can be useful in the context of web-based projects, where numerous concepts such as standards [20], web accessibility guidelines and recommendations [23,24], specific laws for web-based system accessibility [25], etc. must receive attention. These sources of requirements have been formalized in WRM using the *source* element. A catalog of requirements can be extracted from each source. The concepts of catalog and source have some advantages. On the one hand, the stakeholders involved in a web-based development project have at their disposal a repository of the numerous guidelines, laws, etc. involved in the development which, are not always used and, in many occasions, inexperienced practitioners do not know all of them. On the other hand, when a web-based development project is forced to fulfil a law or a guideline, practitioners only need to go to the adequate catalogue and to find in it the requirements that their project must fulfil.

Finally, in WIS scope there is a number of tools that are widely used to check the compliance of the developed application to a given set of predefined rules. Among them, usability and accessibility validators (see e.g. [26,27]) are very popular. The inclusion in WRM of the *external tool* concept together with its relationship with a *catalog* that gathers the rules that the tool contemplates eases the automation of the evaluation process. With these relationships, the practitioners involved in the project can know the tools existing for accessibility and usability validation and what requirements can be validated using them.

4   Applying the Proposed Requirements Metamodel

This section illustrates how WRM can be used for the elicitation of requirements using a study case corresponding to a simplified on-line ticket sale system.

4.1   Study Case Description

For the sake of simplicity, let's assume that the *CEO* of a certain cinema chain wants to *increase the sales profit* for the company. With this aim, let's also suppose that this goal can be achieved through two different sub-goals. On one hand the company wants to increase the sales net profit by implementing an on-line ticket sale system that decreases the costs associated with the sales process. On the other hand, the company also wants to *increase the number of sales* by reaching a broader range of customers. The CEO believes that offering the tickets through the Internet may positively influence both sub-goals, so s/he has embarked into a web ticket sales system development process.

As for the *web customer* that is going to interact with the application, s/he has as the main functional requirement that of *buy ticket*. This requirement can be decomposed into two functional requirements: *browse available films* and *purchase ticket*. In addition, several non-functional requirements that the web based system must fulfil have been identified. Firstly, the *buy ticket* functionality should follow *accessibility* guidelines to allow web customers with disabilities to access the system. Additionally, the system should provide *information accuracy* while browsing through the available films: sessions, prices and so forth should be reliable. Also, the application *learnability* should be high, that is, the application should be simple enough for novel users to easily learn its operation. Last, the purchase process should be performed assuring the *security* of the customer data.

4.2 Instantiating the WRM Metamodel

If we check the involved WRM metaclasses (see Fig.1), we can observe how: (i) the sales manager and web customer both instantiate the *Stakeholder* metaclass, (ii) all goals related to increasing sales profit instantiate the *Goal* metaclass, (iii) buy tickets, browse available films, purchase tickets are all *Functional Requirement* occurrences and (iv) acessibility, learnability, security and information accuracy are all *Non Functional Requirement* instances. Logically, these requirements are decomposed in other more concrete which has not been shown for the sake of simplicity. WRM can help to the systematic identification of these requirements which serves modelers as a good baseline to start the WIS modelling. These identified requirements must be reflected on WIS models, such as content, navigation or presentation models.

5 Automatic Support for WRM

Once WRM was defined, our next aim was the design of a tool that supported it so that stakeholders could manage the WRM concepts in a comfortable way. Next sections explain the considerations made to choose the adequate technological space to develop the tool and the appearance and features of the first prototype implemented to study the feasibility of the approach.

5.1 Technological Environment

For the selection of the technological environment, some considerations had to be taken into account. First, it was necessary to use an environment that permitted the definition of metamodels in an easy way. Moreover, as we want that different stakeholders (and not only designers) use WRM, we needed to offer them a graphical tool with a usable and comfortable interface that allowed them to create and manipulate models compliant with WRM in an easy way. On the other hand, WRM and its associated tool are not isolated efforts but efforts oriented to their integration in WIS development processes with the aim of reinforcing the requirements management activities. For that, the capacities offered by the technological environment in order to extend the tool with new functionalities or to integrate it in existing tools used in WIS development had to be considered.

Given these premises, the Eclipse platform and, in particular, the Eclipse Modelling Framework (EMF [28]) was selected to implement the tool. Eclipse and

EMF offer some suitable features that make them interesting for our approach. First, EMF offers support to deal with MOF, the standard that OMG recommends for describing metamodels. For that, EMF is useful to describe WRM as well as for the creation and manipulation of models and metamodels. Moreover, it is an open source platform-independent project and their architecture based on plugins makes easy to reuse and to add functionality. Another important advantage is that some tools used for WIS development [29] are being migrated to this platform and some of the efforts [7] to homogenize the concepts managed in WE use Eclipse as technological environment. The integration of our automatic support in these Eclipse tools can be directly obtained.

5.2 Eclipse Tool Support for WRM

The first step for the creation of the tool that supports the approach was the formal definition of WRM using EMF. This definition allows the stakeholders to manipulate WRM concepts and to design requirements models compliant with WRM using a tree structure editor provided by Eclipse. Even more important, EMF is the basis on which the GMF (Graphical Modelling Framework) of Eclipse can be used to create graphical editors to deal with the concepts defined in the metamodels created with EMF. It is a well-known fact that a graphical editor is more useful, intuitive and comfortable for most of stakeholders and, for that, the implementation of a graphical editor that allows modellers to deal with WRM concepts in an easy way was tackled. The appearance of this graphical editor is presented in Figure 2, which shows a fragment of the requirements model for the study case shown in Section 4, where some elements are missed for lack of space. In this Figure extra descriptions have been added to easy its understanding.

On the right side, the tool shows a palette to manage the concepts in WRM next to an intuitive icon. In the example shown in Figure 2, the CEO has pro-

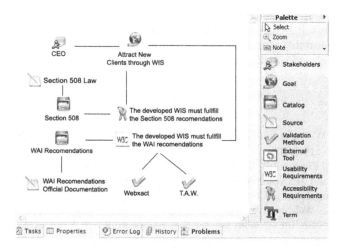

Fig. 2. WRM automatic support

posed a new goal: to attract new clients through a WIS for on-line ticket sale. Once this goal has been defined, modellers can define a set of requirements to fulfil it. In the Figure 2 we assume that up to then, just one usability and one accessibility requirement have been identified. Of course, the complete model would include more requirements, as well as other elements. The accessibility requirement establishes that the WIS must fulfil the WAI recommendations. If exists, we can reuse the catalogue that gathers the requirements extracted from the official WAI documentation. New requirements on the other hand would need to be completely specified and added to the catalogue, and from that time on they would become available for any further project. Moreover, two validation tools have been associated to ensure the fulfilment of this requirement.

6   Conclusions and Further Work

The adequate management of requirements is a key factor for the software systems stakeholders satisfaction. However, neither organizations proprietary development processes nor WIS development methodologies have so far succeeded in tackling this issue. This fact is special worrisomely with quality requirements such as usability or accessibility, which do not receive the attention that they deserve in the first stages of WIS developing proccess. WRM offers support for a structured requirements management. Its design has been done so that it can be used in WIS development processes and can fill the gap detected in WIS methodologies with regard to requirements management. Moreover, the presented automatic support allows stakeholders to deal with project requirements in an easy way. As it occurs with non web-based systems, we believe that the efforts invested in an adequate management of the activities related to Requirements Engineering will pay off in the shape of a reduction of failures in the system and an increased WIS quality, both of them necessary features in our effort to increase the end user satisfaction.

Some further lines of research include the definition of transformation rules from web requirements modelled with WRM and WIS-related development artifacts and, related to this, the explicit treatment of requirements traceability.

References

1. Lowe, D.: Web system requirements: an overview. Requirements Engineering Journal 8, 102–113 (2003)
2. Lang, M., Fitzgerald, B.: Web-based systems design: a study of contemporary practices and an explanatory framework based on method-in-action. Requirements Engineering Journal 12, 203–220 (2007)
3. Ginige, A.: Web engineering: Managing the complexity of web systems development. In: 14th Int. Conference on SW Engineering and Knowledge Engineering (2002)
4. Epner, M.: Poor project management number-one problem of outsourced e-projects. Research Briefs, Cutter Consortium (2000)

5. Glass, R.: Software Engineering: Facts and Fallacies. Addison-Wesley, Reading (2002)
6. Escalona, M.J., Koch, N.: Requirements engineering for web applications: A comparative study. Journal of Web Engineering N 3, 193–212 (2004)
7. MDWEnet-Initiative: Web Engineering Metamodel (2008), http://www.big.tuwien.ac.at/projects/mdwenet/
8. Blaine, J.D., Cleland-Huang, J.: Software quality requirements: How to balance competing priorities. IEEE Software, 22–24 (2008)
9. Damian, D., Chisan, J.: An empirical study of the complex relationships between requirements engineering processes and other processes that lead to payoffs in productivity, quality, and risk management. IEEE Transactions on Software Engineering 32, 433–453 (2006)
10. Kleppe, A.G., Warmer, J., Bast, W.: MDA Explained: The Model Driven Architecture: Practice and Promise. Addison-Wesley Publishing Co., USA (2003)
11. Koch, N., Zhang, G., Cuaresma, M.J.E.: Model transformations from requirements to web system design. In: ICWE, pp. 281–288 (2006)
12. Ceri, S., Fraternali, P., Bongio, A., Brambilla, M., Comai, S., Matera, M.: Designing Data-Intensive Web Applications. Morgan Kaufmann Publishers Inc., San Francisco (2002)
13. Koch, N., Kraus, A.: The expressive power of uml-based web engineering. In: 2nd Int. Worskhop on Web-oriented Software Technology (IWWOST) (2002)
14. Gómez, J., Cachero, C.: OO-H: Extending UML to Model Web Interfaces. Information Modelling for Internet Applications. IDEA Group Publishing (2002)
15. Cuaresma, M.J.E., Koch, N.: Metamodeling the requirements of web systems. In: WEBIST (Selected Papers), pp. 267–280 (2006)
16. Bolchini, D., Paolini, P.: Goal-driven requirements analysis for hypermedia-intensive web applications. Requirements Engineering 9, 85–103 (2004)
17. Yu, E.S.: Modeling organizations for information systems requirements engineering. In: Proc. 1st IEEE Int. Symposium on RE, pp. 34–41 (1993)
18. Berre, A.J.: Comet (component and model based development methodology) (2006), http://modelbased.net/comet/
19. Vicente-Chicote, C., Moros, B., Toval, A.: Remm-studio: an integrated model-driven environment for requirements specification, validation and formatting. Journal of Object Technology, Special Issue TOOLS EUROPE 2007 6, 437–454 (2007)
20. I.S.O.: Iso/iec 9126-1. softw. eng.-product quality - part 1: Quality model (2000)
21. Escalona, M.J., Torres, J., Mejías, M.: Requirements capture workflow in global information systems. In: OOIS, pp. 267–279 (2002)
22. Toval, A., et al.: Eight key issues for an effective reuse. International Journal of Computer Systems Science. Accepted for publication (2007)
23. (W3C), W.W.W.C.: (Web accessibility initiative (wai), http://www.w3.org/
24. Nielsen, J.: Designing Web Usability: The practice of Simplicity. New Riders Publishing (2000)
25. Government, U.: Section 508: The road to accessibility (2008), http://www.section508.gov/
26. Abascal, J., Arrue, M., Garay, N., Tomás, J.: Evaliris: A web service for web accessibility evaluation. In: 12th Int. WWW Conference, Budapest, Hungary (2003)
27. Watchfire: Webxact (May 2008), http://webxact.watchfire.com/
28. Project, E.M.F.: Eclipse modeling framework (emf) (2008), http://www.eclipse.org/modeling/emf/
29. Acerbis, R., Bongio, A., Brambilla, M., Butti, S.: Webratio 5: An eclipse-based case tool for engineering web applications. In: ICWE, pp. 501–505 (2007)

Towards an Ontology-Based Approach for Dealing with Web Guidelines

Joseph Xiong, Christelle Farenc, and Marco Winckler

IHCS-IRIT, Université Paul Sabatier,
118 route de Narbonne, 31062 Toulouse Cedex 4, France
{xiong,farenc,winckler}@irit.fr

Abstract. This paper presents an Ontology-based approach for dealing with guidelines concerning the usability and the accessibility of Web applications. We report an ontology which provides a formal description of concepts used to express ergonomic knowledge related to Web design. This paper demonstrates how to employ such Ontology for organizing ergonomic knowledge by the means of guidelines and, in particular how to employ such guidelines during the inspection of Web-based user interfaces.

Keywords: User interface, ergonomic knowledge, Ontology, accessibility, guidelines inspection, web applications.

1 Introduction

Usability and Accessibility are widely recognized as important requirements for user acceptance of interactive systems. These requirements become even more critical due to the hard concurrency between Web sites (e.g. electronic commerce applications [18]) and legal commitment for quality of information delivery (e.g. Accessibility responsibility of content published on the Web) [23].

Most of the currently available knowledge concerning ergonomic Web-based user interfaces is currently available by the means of guidelines which may cover Usability and Accessibility issues. These guidelines are largely available in many different formats with contents varying both in quality and level of detail [10, 14]. Many studies have shown that careful application of guidelines had positive impact on usability [13, 18]. However, the use of guidelines is not straightforward for developers and evaluators. On one hand, guidelines are quite often described in natural language so that they must be interpreted before to be properly applied to the user interface. On the other hand, the availability of large amount of guidelines sources makes difficult to identify those guidelines which better address the problems for a particular Web site. These problems have lead, ultimately, to the development of specialized tools for organizing guidelines [11, 19], authoring tools for assisting users (i.e. designers and authors) to provide accessible and usable content [16, 17] and tools supporting automated guidelines inspection [1, 2, 20].

Several tools supporting automatic inspection of the HTML/CSS code of Web pages emerged as a valuable approach for improving usability of Web applications [3]. Such tools are easy to use even for non-experts; however they have some

S. Hartmann et al. (Eds.): WISE 2008, LNCS 5176, pp. 132–141, 2008.

limitations. On one hand, these tools only operate on fixed elements HTML tags and attributes thus reducing the scope of the inspection to user interface elements which are only available after the Web applications has been implemented. By doing so, only guidelines that can be translated to HTML tags and attributes are inspected, and quite often part of the semantics of guidelines is lost during this translation. On the other hand, rule engines and guidelines are tight coupled by hard-coded algorithms automating the inspection of user interface elements. So that it's very difficult to modify existing guidelines or to introduce new ones into these tools. The cost involved on the introduction of new guidelines might prevent the evolution of these tools towards the inspection of other artifacts (e.g. models produced according to Model Driven Engineering and Web Engineering approaches [4]) and/or new technologies used to build Web applications such as AJAX [9]. It is noteworthy that most guidelines do not refer to HTML/CSS elements/attributes or any other particular technology.

Behind all these concerns there is a mapping-problem between the way guidelines are usually described and how they can be applied during the design and evaluation. Our main assumption is that an Ontology-based approach could be useful to overcome such mapping-problems by providing a formal and non-ambiguous vocabulary. For that, in this paper we introduce an Ontology for dealing with usability and accessibility guidelines for Web applications. The rest of the paper is organized as follows: section 2 introduces an Ontology-based approach for organizing and encoding usability guidelines using such Ontology. Section 3 describes the organization of guidelines. Section 4 presents a general discussion concerning the use of such usability guidelines at different phases of the development process. Lately, section 5 presents a discussion and final remarks.

2   An Ontology-Based Approach for Encoding Usability Guidelines

Currently there are several guidelines sources, some of them devoted to usability [18, 19] and/or accessibility [21]. Sometimes guidelines compilations are presented according the application domain, for example: Graphical User Interface guidelines, Web guidelines, mobile guidelines and so on. For the purposes of this paper, we designated by the term guideline any recommendation that could be applied to user interface of Web applications. Guidelines are expressed in natural language and it leads to several difficulties: ambiguities due to the natural language, guidelines expressed as general statement (no explicit indications on what should be evaluated), misunderstanding of the actual meaning due to the loss of context, etc. For instance, the guideline *"Test the navigation design"* is too imprecise and must be interpreted. One possible interpretation of this guideline is *"(to) avoid pages that contain dead links and pages that contain no link"*. Thus, before evaluating a guideline one may give an interpretation when needed. However, it might have different interpretations of the actual meaning of one guideline sometimes due to the difference of their level of expertise. This problem is inherent to guidelines and evaluators will always be confronted to this difficulty. However, a standard vocabulary could provide a formal and non-ambiguous description of concepts embedded into guidelines. Once this vocabulary is well established guidelines can be written in a declarative form using the terms contained in the vocabulary.

An ontology is aimed to empower the semantic of terms related to one domain and their relationships and thus could establish this vocabulary. Furthermore, it helps to uniform guidelines statements. For example, guidelines *"Include on the navigation aid pointers across to main sections"* and *"Provide persistent links to the home page and high-level site categories"* mention the terms *"pointers"* and *"links"* that refer to the same concept. An appropriate ontology will be helpful by unifying these two terms in a single term, for instance, the term "Link". In addition, the ontology establishes a non-ambiguous semantic for what the term "link" means.

After revising the W3C/WAI Content Accessibility Guidelines 1.0 [21] and the set of guidelines compiled during the EvalWeb project [15] (which covered a comprehensive number of guidelines sources), we have extracted and formally described the list of concepts presented in Fig. 1.

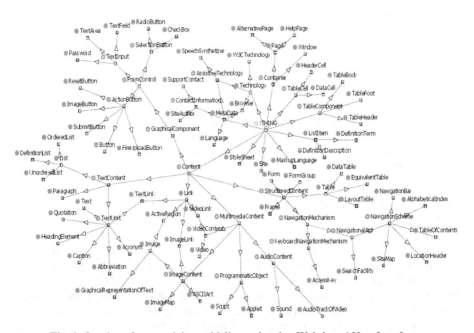

Fig. 1. Ontology for organizing guidelines related to Web-based User Interfaces

This lightweight Ontology, as defined by Corcho et al [6], does not include any rule for inferring knowledge. The formal definition of concepts is coded using the Ontology Web Language (OWL) [12]. The current version of such ontology defines 94 concepts such as *"Page"*, *"Link"*, *"Table"*, *"Frame"*, *"NavigationalAid"*, *"SiteMap"*, etc. which are organized around four main categories, namely: *site, container, page* and *content*. Each concept has a specific semantic meaning and a list of attributes that can be used during both manual and automated inspection of guidelines on Web-based user interfaces. Fig. 2 presents a class diagram for the concept *"Page"*. Basically, a *"Page"* element represents a simple Web page. More specialized pages

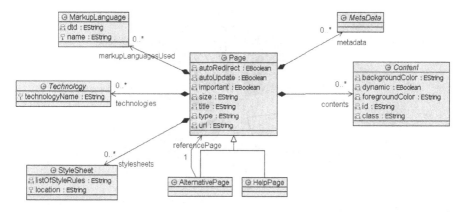

Fig. 2. Class diagram and list of attributes for the concept *"Page"*

can be alternative pages or help pages. One Web page can mention the technologies and/or markup languages used. Besides, style sheets, metadata elements and contents can be attached to this page.

Table 1 presents the list of corresponding attributes associated to this concept. The complete ECore version of all concepts of the ontology is available at http://ihcs. irit.fr/xiong/ontology.html.

Table 1. List of attributes associated to the concept *"Page"*

Attribute name	Attribute type	Description
autoRedirect	boolean	Indicates if this page automatically redirects the user to another page
autoUpdate	boolean	Indicates if this page performs an automatic update at specific intervals
important	boolean	Indicates if this page is important or not
size	string	The size of this page (e.g. in Kb)
title	string	The title of this page
type	string	The role this page plays in the Web site (e.g. "homepage")
url	string	The URL of this page.
contents	List(Content)	The list of contents of this page.
markupLanguagesUsed	List(MarkupLanguage)	The list of markup languages used in this page.
metadata	List(MetaData)	The set of metadata attached to this page.
stylesheets	List(StyleSheet)	The list of style sheets attached to this page.
technologies	List(Technology)	The list of technologies used in this page.

The concepts described by the Ontology only cover elements addressed by a given corpus of ergonomic knowledge. So it would be possible to have User Interface elements that are not part of this set of concepts because there are no guidelines addressing them. In order to exemplify this situation, Fig. 3 presents the distributions of concepts as they are addressed by W3C/WAI Content Accessibility Guidelines 1.0.

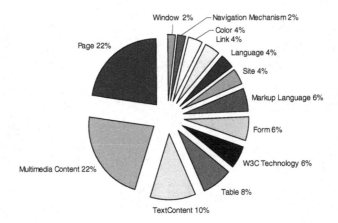

Fig. 3. Distribution of Ontology concepts according to the W3C/WAI guidelines

3   Guidelines Encoding and Evaluation

As part of the ontology definition there is a list of operations that could be applied on each concept. These operations are used as conditions on guidelines thus allowing the detection of violations. Fig. 4 shows part of the list of operations we defined. Each operation on ontology terms is hard-coded using the concepts definitions and attributes using the definitions given by the set of guidelines. For example, the body of the operation HasTextualAlternative is as follows (in a Java-like syntax): The body of this operation only involves a variable of type MultimediaContent which is a term of the ontology. Thus, the evaluation logic is embedded in these operations.

```
Boolean HasTextualAlternative (MultimediaContent content) {
       return content.getTextualAlternative() != null;
}
```

```
HasTextualAlternative              :MultimediaContent   → Boolean
HasLongDescription                 :MultimediaContent   → Boolean
HasSufficientContrastedColors      :Content             → Boolean
IsLinkedToHomePage                 :Page                → Boolean
IsEmptyString                      :String              → Boolean
ContainsHeaders                    :String              → Boolean
ContainsRelativeUnits              :StyleSheet          → Boolean
ExistsBackgroundAndForegroundColors :StyleSheet         → Boolean
IsNull                             :Thing               → Boolean
...
```

Fig. 4. Excerpt of operations on ontology terms

Let's assume that, from now, guidelines that are ambiguous or imprecise are interpreted so as to be suitable for evaluation. So that guidelines should be rewritten according to the terms used to express ontology concepts. Let us consider the two guidelines: The first guideline is (G1) *"Each page should contain at least one link"* and the second one (G2) *"Ensure that foreground and background color combinations provide sufficient contrast"*. These guidelines involve the identification of the following terms: page and link (G1); and foreground color and background color (G2). In our ontology these terms correspond respectively to the concepts *"Page"* and *"Link"* and to the attributes *"foregroundColor"* and *"backgroundColor"* of the term *"Content"*. The operations involved in these guidelines are respectively *"ContainsLinks"* and *"HasSufficientContrastedColors"*. Thus, G1 and G2 can be respectively described as:

g1: ContainsLinks(page)
g2: HasSufficientContrastedColors(content)

Where:
g1 and g2 stands for guideline 1 and guideline 2;
page is of type Page and content is of type Content (Page and Content are two terms of the ontology).

Although G1 involves an object of type Link it does not appear in g1. Actually, G1 does involve an object of type Link but it is embedded in the body of the operation ContainsLinks which checks whether this page has at least one link or not. We can have the same remark concerning the variable named content: it does not mention the terms ForegroundColor and BackgroundColor. But in this case it is more evident as they are two attributes of content.

So far guidelines can be encoded as operation on terms of the Ontology. Thus, given a set of encoded guidelines and a Web application, it is possible to formalize the evaluation process as follow:

Eval(**WebApp**, **g**) = Exec(Map(**WebApp**,o), **g**) = {"Respected", "Error", "Warning", "Non Verifiable"}

Where:
Eval is the evaluation function that returns the result of the evaluation: *Respected, Error, Warning, Non Verifiable;*
Exec is the evaluation function that takes as input ontology terms and a guideline;
Map is the mapping function that maps the Web application terms to the Ontology terms;
WebApp is the Web application;
$g^i \in G$ is the set of encoded guidelines;
$o^i \in O$ is the set of concepts in the Ontology

The evaluation of a guideline consists in two phases: a mapping phase and a validation phase. In the mapping phase information is collected from the evaluated Web application in order to build a representation using terms of the ontology. For example, if we parse an HTML file, each <img> tag will correspond to the Image term in the ontology and its alt attribute will correspond to the Image's textualAlternative attribute, and so on for each term of the ontology.

Once this mapping has been done the validation phase can take place (this is the role of the Exec function). In this phase we have a set of instantiated ontology terms that represents the values collected from the evaluated Web application in the previous mapping phase. These values are then used to determine whether all conditions of the guideline are satisfied or not. Conditions, i.e. operations on terms, range from simple conditions such as testing attributes of a term, to more complex conditions such as verifying that the color contrast is sufficient. We note that when evaluating a non-leaf term, i.e. a term that has sub-terms in the ontology, it also includes the lower terms in the hierarchy. For example, if a guideline mentions the NavigationalAid term (see Fig.1) it addresses all navigation schemes, i.e. alphabetical indexes, location headers, navigation bars, site maps and tables of content.

The result of the evaluation, i.e. the result of the Exec function, depends on the condition satisfaction. When the condition is satisfied the result is "Respected". When a condition is not satisfied an "Error" or "Warning" is thrown. When one guideline cannot be automatically verified the result is "Non Verifiable".

4   Generalizing Guidelines Inspection to Other Web Artifacts

Most guidelines do not refer to HTML elements/attributes or any other particular technology. Neither they are limited to the inspection of advanced prototypes. In fact, according to the phase of the development life cycle different categories of guidelines can be employed for supporting design and/or evaluation of User Interfaces (UIs) [19]. For example, during early design phase, guidelines inspection can be performed over templates based on mockups or sketches. The early identification of usability problems over such templates could reduce the time and the costs of automated in-spections of pages created from the template. In addition, some guidelines, for exam-ple *"Provide enough color contrast between foreground and background"* could be applied to many different artifacts produced during the design processes (such as templates and HTML pages).

According to design choices, strategies and constraints, Web applications can be modeled in many ways (navigation model, structure model, etc.) and implemented in many languages (HTML, PHP, XML/XSL, etc.). Consequently, this leads to com-plexity in the task of evaluation. Hence, the challenge is to be capable of evaluating Web applications whatever the model or technology used. To overcome this issue we use our set of guidelines (formally described with the ontology terms) in combination with mapping tables. Mapping tables are the means by which we link terms in the ontology with its corresponding one in the target artifact. Thus, for one artifact we provide its mapping table that contains each mapping. This mapping table can be assimilated to a mapping function. Fig. 5 schematizes how we can exploit guidelines to evaluate different artifacts.

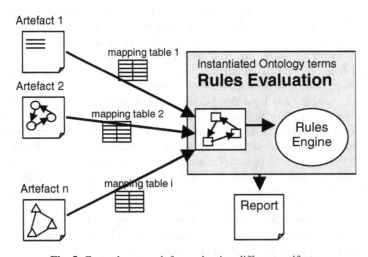

Fig. 5. General approach for evaluation different artifacts

The responsibility of the mapping table is to identify each term involved in a guideline on one target artifact (e.g. what does the term *"Link"* refer to on HTML pages?). This will be useful to create an instance of the Web application using instantiated ontology terms, i.e. a set of ontology terms in which attributes are instantiated are created and represents the real application. This allows us to identify the objects to evaluate before the evaluation takes place. As each artifact comes with its mapping table for *n* artifacts there is *m* mapping functions, which can be formally described as follows:

MapHTML(WebApplicationHTML, O) = {t}HTML
MapXML(WebApplicationXML, O) = {t}XML
MapJavaScript(WebApplicationJavaScript,O) = {t}JavaScript
...

Where:
Map^i is the mapping function for artifact i;
WebApplication^i is the Web application developed/modeled in language i;
O is the ontology of terms for Web applications;
{t}i is the set of instantiated ontology terms (as result of the mapping function).

Let us give an example of mappings with the previous guidelines (G1) *"Each page should contain at least one link"* and (G2) *"Ensure that foreground and background color combinations provide sufficient contrast"*. Assume that we are evaluating a navigation model of one Web application (in this example we use state diagrams to model navigation) and the same application implemented in the HTML language.

Table 2. Examples of mappings between Web artifacts and Ontology concepts

Guideline	Ontology concepts and attributes	State diagrams	HTML tags and attributes
G1	Page	State	\<html\>
	Link	Transition	\<a\>
G2	Background color	-	bgcolor
	Foreground color	-	color

Guidelines G1 and G2 involve the identification of 4 terms from the ontology: Page, Link, Background color, and Foreground color. Table 1 shows the mappings between ontology terms and artifacts terms. When using state diagrams, pages are represented by states and links by transitions. However, there is no possible mapping with the terms Background color and Foreground color as they can't be represented with state diagrams. In HTML, pages and links are respectively represented with the \<html\> and \<a\> tags. Concerning Background color and Foreground color, they can be mapped with the *bgcolor* and *color* attributes of HTML elements.

As shown in Table 2, it is not always possible to establish mappings. This is due to the fact that some artifacts are not expressive enough to represent all concepts. For example, it is impossible to describe background color and foreground color on state diagrams. When a guideline involves an ontology term that has no representation in a given artifact this guideline cannot be automatically evaluated. The approach does not impose the use of any particular methodology for developing Web application. On one hand it can be adapted to any kind of artifact used during the development process. One other hand, however, it depends on the expressiveness of the notations employed.

5 Discussion and Final Remarks

This work introduced an ontology-based approach for organizing guidelines related to the development of Web-based user interfaces. Such descriptions of guidelines require a compromise between the system expert (which knows all artifacts and models used to build Web applications) and an ergonomics expert (who systematizes ergonomic knowledge by the means of guidelines). The general goal is similar to other model-based approaches for developing quality Web sites [4, 5]; however, the specificity of the current work is to generalize guidelines from an existing corpus of ergonomic knowledge. Whilst this ontology is based on sound sources of ergonomic knowledge, it has not yet been validated by stakeholders. For the moment, we assume that the common vocabulary it provides is useful to demonstrate the practical application of an ontology-based approach for inspecting Web-base user interfaces.

The availability of guidelines encoded according to the Ontology provides a powerful support for generalizing the approach to many different artifacts produced during the life cycle of Web applications. Due to space reasons, the description of the guidelines is limited to a few examples of W3C/WAI guidelines. However, the contribution must not be understood as limited to this set of accessibility guidelines. The implementation of the corresponding mapping between guidelines and artifacts is very costly, as we have experienced during the creation of mappings to HTML/CSS code and SWC models [22]. However, once we have established the mapping tables to a given artifact users can benefit from it for all new projects, thus reducing the costs over time.

One of the main drawbacks of our approach is that designers should specify all elements of the UIs at different levels and if they miss to clearly identify an element, it will not be checked by our tools. This problem can be alleviated by appropriate authoring environments supporting the design and implementation of Web sites.

Future work will include the extension of such an approach for other artifacts produced during the development process. In addition, we have planned the validation of the Ontology with stakeholders and a detailed comparison of on such ontology-based approach with currently available tools for automating the evaluation of Web applications.

Last but not least, these preliminary results have been applied to an industrial project e-Citiz [7] and actually are fully supported by an Eclipse plug-in.

Acknowledgements

This research is supported by the action COST294-MAUSE (http://cost294.org), ANRT-CIFRE and Genigraph (http://www.genigraph.fr), Genitech Group.

References

1. A-Prompt, Web Accessibility Verifier, http://aprompt.snow.utoronto.ca/
2. Beirekdar, A., Vanderdonckt, J.: Noirhomme-Fraiture, M. A Framework and a Language for Usability Automatic Evaluation of Web Sites by Static Analysis of HTML Source Code. In: Proc. of CADUI 2002, pp. 337–348. Kluwer Academics Pub., Dordrecht (2002)
3. Brajnik, G.: Automatic Web Usability Evaluation: What Needs to be Done? In: Proc. of 6th Conf. on Human Factors and the Web HF Web 2000, Austin, USA, 19 June (2000)
4. Cachero, C., Poels, G., Calero, C.: Metamodeling the Quality of the Web Development Process' Intermediate Artifacts. In: Baresi, L., Fraternali, P., Houben, G.-J. (eds.) ICWE 2007. LNCS, vol. 4607, pp. 74–89. Springer, Heidelberg (2007)

5. Centeno, V., Kloos, C., del Toro, J.M.B., Gaedke, M.: Web Accessibility Evaluation Via XSLT. In: WISE Workshops 2007, pp. 459–469 (2007)
6. Corcho, O., Fernández-López, M., Gómez-Pérez, A.: Methodologies, tools and languages for building ontologies. Where is their meeting point? Data and Knowledge Engineering 46(1), 41–64 (2002)
7. e-Citiz, Le monde des téléservices à portée de clic, http://www.e-citiz.com/
8. Gaedke, M., Lowe, D. (eds.): Web Engineering- 5th international conference, ICWE 2005, Sydney, Australia, July 27-29 (2005)
9. Garrett, J.A.: A New Approach to Web Applications (February 2005), http://www.adaptivepath.com/publications/essays/archives/000385.php
10. Mariage, C., Vanderdonckt, J., Pribeanu, C.: State of the Art of Web Usability Guidelines. In: Proctor, R.W., Vu, K.-Ph.L. (eds.) The Handbook of Human Factors in Web Design, ch. 41. Lawrence Erlbaum Associates, Mahwah (2005)
11. Mariage, C., Chevalier, A., Vanderdonckt, J.: Using the MetroWeb Tool to Improve Usability Quality of Web Sites. In: Proc. of LA-Web 2005 Conference, Buenos Aires, 31 October - 2nd November 2005). IEEE CS Press, Los Alamitos (2005)
12. OWL Web Ontology Language: Overview. W3C Recommendation (February 10, 2004), http://www.w3.org/TR/owl-features/
13. Ratner, J., Grose, E., Forsythe, C.: Characterization and Assessment of HTML Style Guides. In: Proc. of ACM CHI 1996, vol. 2, pp. 115–116 (1996)
14. Scapin, D., Kortum, P., et al.: Towards automated testing of web usability guidelines. In: Kortum, Ph., Kudzinger, E. (eds.) Proc. of 6th Conf. on Human Factors and the Web HFWeb 2000, 19 June 2000. University of Texas, Austin (2000)
15. Scapin, D., et al.: Conception ergonomique d'interfaces web: démarche et outil logiciel de guidage et de support, INRIA Technical Report of EvalWeb project, INRIA-Université de Toulouse 1-UCL, Rocquencourt-Toulouse-Louvain (December 1999)
16. Theng, Y.L., Rigny, C., Thimbleby, H., Jones, M.: HyperAT: HCI and Web Authoring. In: O'Conaill, Thomas, P.J. (eds.) Proceedings of People and Computers XII, Bristol, pp. 359–378 (August 1997)
17. Treviranus, J., McCathieNevile, C., Jacobs, I., Richards, I. (eds.): Authoring Tool Accessibility Guidelines 1.0. W3C Recommendation 3 (February 2000), http://www.w3.org/TR/WAI-AUTOOLS/
18. van Duyne, D.K., Landay, J.A., Hong, J.I.: The Design of Sites: Patterns, Principles and Processes for Crafting a Costumer-centered Web experience, pages 762. Addison-Wesley, Boston
19. Vanderdonckt, J.: Development Milestones towards a Tool for Working with Guidelines. Interacting with Computers 12(2), 81–118 (1999)
20. Watchfire WebXACT, http://webxact.watchfire.com/
21. Web Content Accessibility Guidelines 1.0. W3C Recommendation (5-May-1999), http://www.w3.org/TR/WAI-WEBCONTENT/
22. Winckler, M., Palanque, P.: StateWebCharts: a Formal Description Technique Dedicated to Navigation Modelling of Web Applications, DSV-IS, Portugal (2003)
23. Winckler, M., Xiong, J., Noirhomme-Fraiture, M.: Accessibility Legislation and Codes of Practice: an Accessibility Study of Web Sites of French and Belgium Local Administrations. In: proceedings of the 1st International Workshop on Design & Evaluation of e-Government Applications and Services (DEGAS 2007), Rio de Janeiro, Brazil (September 11th, 2007); CEUR Workshop Proceedings (ISSN 1613-0073), http://ceur-ws.org/Vol-285

MEM&LCW 2008 Workshop PC Chairs' Message

Marek Kowalkiewicz[1], Dominik Flejter[2],
and Tomasz Kaczmarek[2]

[1] SAP Research Brisbane, Australia
[2] Poznan University of Economics, Poland

Today's business world relies on a range of sophisticated IT solutions. Due to the sophistication, business and IT have separated in many organizations. That separation has left a gap between the two and has led to frequent IT project failures or escalations. While the gap between business and IT does not pose a difficulty for highly structured business problems and processes that are stable in time, it is a major obstacle for all other problems and processes.

Web 2.0 technologies and paradigms, such as mashups and lightweight composition, are believed to provide a handle to overcome this situation and enable a potential candidate for a solution that will be essential in bridging the gap between business and IT. The new technologies, such as mashups, enterprise mashups, and lightweight composition, provide business users with ability to compose and use simple applications that follow their requirements, and not require programming skills, or even understanding of the concept of SOA, while providing the requested functionalities.

The goal of the MEM&LCW 2008 workshop is to provide a platform for discussing research topics underlying the concepts of lightweight composition, mashups, and enterprise mashups. By bringing together representatives of academia and industry, the workshop is also an important venue for identifying new research problems and disseminating results of the research. By affiliating with a renowned international conference, the workshop provides a possibility to interact with researchers from other areas of the domain of Information Systems.

This year's edition of the workshop has attracted a number of excellent submissions from around the world. In order to achieve high quality of the workshop only a small number of submissions was accepted. All of the accepted papers provide a significant contribution to the field and at the same time are highly interesting for enterprises seeking inspirations to help them introduce the new concept in their daily routines.

Four papers are going to be presented during the workshop (ordered alphabetically):

1. A Web based Mashup Platform for Enterprise 2.0 by Rama Gurram, Brian Mo and Ralf Gueldemeister;
2. Bill Organiser Portal: A Case Study on End-User Composition by Agnes Ro, Lily Shu-Yi Xia, Hye-young Paik and Chea Hyon Chon;

S. Hartmann et al. (Eds.): WISE 2008, LNCS 5176, pp. 142–143, 2008.

3. Extending Services Delivery with Lightweight Composition by Christian Janiesch, Kathrin Fleischmann and Alexander Dreiling;
4. Fixed-mobile Hybrid Mashups: Applying the REST Principles to Mobile-specific Resources by Sami Mäkeläinen and Timo Alakoski.

Rama Gurram, Brian Mo and Ralf Gueldemeister from SAP Labs in Palo Alto present Enterprise Mashup Application Platform - a browser based application for composing widgets into mashup applications. It simplifies development of mashups on top of Enterprise Systems by providing means to bind together enterprise data sources, and existing widgets. The widgets communicate with each other via events and are able to consume news feeds or connect to RESTful services.

Ro, Xia, Paik and Chon introduce a lightweight end-user service composition paradigm, based on three atomic concepts: Stones, Stories, and Story Boards. They show how that concept can be used in constructing new mashups. Initial studies of user acceptance show very positive results.

Christian Janiesch, Kathrin Fleischmann and Alexander Dreiling propose the amalgam of Texo (platform to supply fine grained business services) and RoofTop (AJAX-based lightweight composition platform). It allows to create simple widget-like applications and extend the idea of service delivery by placing the result of service execution in a broader context of other mashed-up information services.

Sami Mäkeläinen and Timo Alakoski discuss the solution (and provide a prototype) to wrap services and resources specific to mobile platforms as RESTful services and enable their composition. They managed successfully to demonstrate that it is possible to design services in a RESTful manner based on the systems built with very different paradigms.

We believe that the excellent submissions will provide a starting point for fruitful discussions during the workshop. We also invite the authors, workshop participants and others to consider active participation in future editions of the workshop.

A Web Based Mashup Platform for Enterprise 2.0

Rama Gurram, Brian Mo, and Ralf Gueldemeister

SAP Labs LLC, 3410 Hillview Ave,
Palo Alto, CA, USA
{rama.gurram,brian.mo,ralf.gueldemeister}@sap.com

Abstract. Traditionally, enterprise applications are complex and monolithic. Business operation has become one of the most important factors to determine the efficiency of an organization. Therefore, great amount of customization has to be made with enterprise applications. An organization will have a big competitive advantage if it can adopt new business requirements and processes faster. Today, customization and extension of enterprise applications are hard, time consuming, and require expert knowledge. By means of extending mashup concepts which are popular in consumer space into enterprise applications space, it is possible to make an enterprise platform friendlier for a new generation of developers and users.

Enterprise Mashup Application Platform (EMAP) is a browser-based application composition environment and run-time, which simplifies application development on top of existing, complex enterprise platforms and services. The lightweight platform provides tools to realize mashup applications from conceptualization to deployment with a focus on extensibility, open standards and integration of 3rd party components. EMAP allows the user to create faster time to market, Do It Yourself (DIY) applications quickly and helps the users to create a variety of useful enterprise solutions.

Keywords: Web 2.0, RIA, REST, SOA, Lightweight, Enterprise Mashup, Platform, Architecture, Ajax, Widgets, Gadgets, Metadata, OpenAjax, DIY.

1 Introduction

With the advent of Internet technologies, applications have moved from the desktop to static one-way web applications to dynamic, interactive user-centric applications that can mashup data and services from multiple sources as we have seen in many consumer web applications. One early Mashup example is *HousingMaps.com* [1] which combines property listings from Craigslist [2] with Google Maps [3] to make apartment searching easier. A constantly growing number of sophisticated examples can be found at ProgrammableWeb.com [4]. In addition to the term "mashup", "widget" is another term that describes an *individual* mini-app that can run on the desktop, e.g. Yahoo Widgets [5], or can be grouped within a browser to create personalized portals, e.g. iGoogle [6].

While the general Internet users have been enjoying the success and great benefits that Web 2.0 has brought, the enterprise space has been lagging behind in the adaptation of Web 2.0 concepts and technologies; IBM's QEDWiki is one of the available

S. Hartmann et al. (Eds.): WISE 2008, LNCS 5176, pp. 144–151, 2008.

solutions [7]. According to a survey from 2007 only very few companies have invested into mashup technologies [8] and a recent paper states important issues for implementing enterprise mashups [9]. In this fast moving business world, the competition is fierce; the question is how companies can serve their business needs in a way that enables fast creation and deployment of personalized business applications which utilize increasingly available data services, internal and external, and reuse components to compose the applications?

2   Shift from Enterprise Portals Towards Enterprise Mashups

In the enterprise space, some companies have offered Web 2.0-like Enterprise Portals to run their web applications. Some of them allow developers to compose enterprise web applications using layout templates that combine HTML, scripting languages and special portal components, i.e. Portlets. Basically, Portlets are mini programs that communicate with the underlying applications and other Portlets, consume and process backend data, and display it on the page.

However many current Enterprise Portals have deficiencies compared to those true Web 2.0 Mashup Portals available in consumer space in terms of application development and rich user interfaces and usability.

Application development with Enterprise Portals is still performed in the traditional way – a rigid and slow process that usually takes months to build and deploy, requiring experienced developers and domain experts. It is desirable to have a lightweight composition platform, which enables rapid development, easy deployment and user specific personalization. Application composition should be metadata-driven and based on open standards, e.g. OpenAjax [10, 11], for reusability and extensibility.

Traditional server-centric Enterprise Portals do not fully utilize the computing power and storage capability of the clients. In addition, communications among Portlets often happens on the server side, thus creating unnecessary round-trip communication between client and server. We need a full-fledged, lightweight client tier that can fully utilize the computing power and storage capability, and enables widgets to communicate directly among each other on the client side. This transition of responsibilities to the client improves the performance and response times.

In traditional Enterprise Portals, there is no easy way to integrate external data or consume external services because they were designed to consume mainly internal data and services from the backend. In order to support mashups natively with data from internal and external resources, radical design changes are required. Many common Web 2.0 data services for mashups use standards such as Extensible Markup Language (XML) or JavaScript Object Notation (JSON) for data exchange via Representational Sate Transfer interfaces (REST).

Although Enterprise Portals have made composition of applications possible, they did not improve the overall user experience. Often, synchronous request-response user interaction patterns are still applied. In contrast, Rich Internet Applications (RIA) can provide a better user experience by following approaches and UI concepts known from desktop application, e.g. drag and drop, flexible layouts and adaptive forms, using technologies like Ajax.

3 Enterprise Mashup Application Platform Prototype

The goal of our research work is to provide an *Enterprise Mashup Application Platform (EMAP)* for effectively composing and integrating mashups in order to make enterprise applications friendlier for a new generation of developers and users. EMAP provides a development and runtime environment that simplifies developing mashup applications on top of current enterprise systems. It features tools to assist developers from conceptualization to deployment and runtime customization to enable a very adaptive user experience.

EMAP provides a simple and lightweight mashup application composition platform to build mashup applications. Reusable widget components are used as building blocks to create enterprise mashup applications by connecting them at the design-time of an application. The interaction between these components is achieved by means of events, which are mediated by the platform on the client side by using a secure subscription based messaging. Usage of metadata allows extensive customization of component behavior and presentation. The metadata syntax and the format of the widget is OpenAjax compliant [12]. Widgets communicate with the enterprise backend through RESTful services or by consuming news feeds to get the data asynchronously either in JSON or in XML format. In addition, EMAP provides a flexible click-n-drag layout of components, support for corporate branding and application state management. Utilizing resource-centric RESTful services for incorporating business data provides the simplicity and enterprise class scalability.

EMAP is accompanied by the Enterprise Widget Repository, which provides a central storage for mashup components. It is fully integrated into the EMAP runtime. The openness of the repository is the foundation for the extensibility of mashup applications. Widget metadata is used for generating the repository's widget directory, including metadata of widgets which do not comply with the OpenAjax Widget standard. Versioning of components helps to realize reliable and recoverable applications.

4 Architecture of an Enterprise Mashup Application Platform

The general architecture of an enterprise mashup application is composed of the following three parts:

Enterprise data sources expose business data through public APIs based on web protocols such as REST, RSS and Atom. Content can be read-only or writable, depending on the data source type and use-case. Access to data sources often follows asynchronous access patterns and open data formats, cf. Ajax, XML and JSON.

Widget Components are reusable building blocks which provide the business and presentation logic on top of data sources. Widgets are enhanced by metadata to enable customization of the provided functionality. Widgets are usually implemented using client-side scripting technologies.

Composite Application Environments provide the design and runtime environment for composing and customizing widgets to realize composite enterprise applications. The environment is responsible for widget layout and client application state management. Existing approaches include web portals such as iGoogle and desktop applications such as Yahoo Widgets.

The EMAP architecture is illustrated in Figure 1. In EMAP, the client's run-time view is same as the design-time view. Mashup application creation, customization and execution can be done in the same environment, based on a user's role.

In EMAP, the application environment is split into four parts, application manager, layout manager, data manager and repository. *Application Manager* provides the widget loading, unloading, life-cycle management, application state management and event exchange mechanism via *Event Hub,* which is compliant with the OpenAjax Hub specification [13]. All widgets definitions and application states are stored in a Repository service. Access to the REST based [14] Repository service is available to the application developer via a design time widget. The Knowledge User can build the lightweight mashup application by dragging required widgets from the Repository widget and connecting the selected widgets by means of pub/sub events. Enterprises can provide domain specific, pre-bundled widgets in the repository to be consumed by mashup applications. Enterprise data sources are available from existing enterprise services or external data providers. The *Data Manager* provides transparent access to the enterprise data along with an optional local storage for offline data access based on application caching policies. *Layout Manager* is responsible for rendering the widgets in a flexible manner along with theme support.

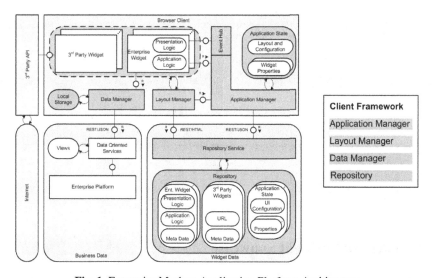

Fig. 1. Enterprise Mashup Application Platform Architecture

The EMAP Programming model is based on creation and composition of reusable components with associated metadata. Knowledge Workers (Domain Experts) compose applications using enterprise widgets or third-party widgets as building blocks and connecting the widgets by means of loosely coupled events. Widgets are created by developers conforming to the standards, i.e. OpenAjax Gadgets, and published in a RESTful Repository using the tools provided by the platform. Proprietary 3rd party widgets, e.g. iGoogle Gadgets, can be used by means of pluggable transformation handlers to the platform. Figure 2 shows a sample OpenAjax-style widget along with metadata. The widget definition contains metadata properties such as layout, display and instance properties, events and embedded or remote content. Instance properties and life-cycle events are exposed to the widget developer through a JavaScript API which can be called from within the widget's business logic implementation.

```
<widget xmlns="http://openajax.org/widget"
    name="BijouxSearchQuick"
    scope="BijouxSearchQuick"
    title="Search"
    height="35"
    resizable="false"
>
    <properties>
        <property name="Connection" type="string"
                value="LDAP">
            <constraint>
                <enum>
                    <option value="LDAP">LDAP</option>
                    <option value="AP">AP</option>
                </enum>
            </constraint>
        </property>
        <property name="contact"
            display_name="Selected Contact Person"
            type="hidden" topic="bijoux.contact_selected"
            publish="true">
        </property>
    </properties>

    <content mode="view" type="frame">
    <![CDATA[
      . .
    ]]>
    </content>
</widget>
```

Fig. 2. Widget Format (OpenAjax style)

5   Contact Management Showcase

A typical enterprise mashup application is illustrated in Figure 3. The application represents the dashboard for a sales user to search and display contact details and acts as the single point-of-entry for this sales user's role. The realized contact management scenario comprises the integration of internal business information as well as external productivity related information.

The application environment enables the composition of pre-defined contact management components and external third-party components. Instances of the first type include the contact search and contact browser components that directly access enterprise data. Instances of the latter type are a to-do list and a weather widget which are imported from third-party component directories and provide access to public services. These two types of runtime widgets are complemented by design-time widgets, which enable the user to add new widgets to the application (Widget Repository) and even create or modify widget definitions (Widget Editor).

This scenario highlights some key features of mashup applications: the integration of external and internal components, dynamic customization by unifying design and run-time environments for the application, and widget composition using pub/sub event based mashups. According to the core philosophy of mashup applications, the user itself can freely change and enhance the application. The user can remove no longer required widgets, add new widgets of interest or rearrange existing widgets to better match his personal workflow and thus tailor the application to very specific needs.

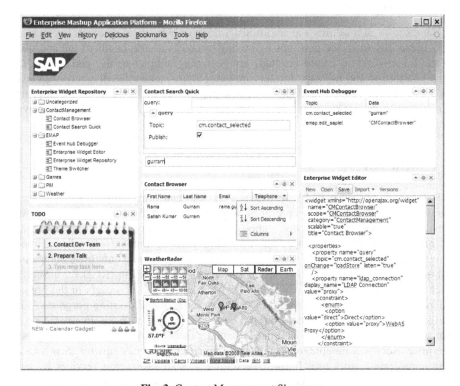

Fig. 3. Contact Management Showcase

6 Key Success Factors

There are several key success factors that influence the adoption of enterprise mash-ups. Some of them have been successfully illustrated by our prototype:

- Rich user experience
- Easy to use tools for development and customization of mashups
- Simple service interfaces to data sources
- Open standards for widget definitions and message formats to ensure exten-sibility of mashups

- Leveraging client-side processing power and reduction of server round trips
- Give the end user the power to innovate

In the context of enterprise application, two important success factors arise which require further research and the integration of enterprise policies. They reflect the higher requirements and compliance issues associated with enterprise solutions:

- Current and reliable data from multiple sources, enabling a confident decision making process
- Security considerations, especially authentication, authorization and secure messaging

The creation of mashups was traditionally done only by the developer, but increasingly the ability to create mashups is moving directly into the hands of the end-user. As the frameworks to create mashups are becoming simpler to use and secure, the widget definition and message formats are being more standardized, enterprise mashups are ready for prime-time. The increased importance of metadata in the development of mashups makes them more flexible in terms of creating end-user driven applications.

7 Conclusion

SAP Research prototype EMAP provides a flexible platform that puts a new generation of knowledge workers in control of What They Want When They Want It. Application developers can quickly compose enterprise-quality applications and users can highly personalize their applications with a growing number of widget components. Because of this flexibility, we predict that more and more enterprise companies will adopt mashup style applications. The Return on Investment (ROI) for a Service Oriented Architecture (SOA) can be realized immediately, when users in the organizations start mixing and matching these services for new and exciting purposes. With a large number of atomic services offered by the IT organization within an enterprise, using REST and SOA based interfaces, which can be easily exposed as widgets – enterprise users will be able to quickly build up situational mashups for their own or their immediate team's use. It can be exciting to start leveraging services by user driven mashups in ways that couldn't have been imagined at the time they were written. As standards, frameworks and tools to build mashups are starting to mature, more enterprises are open to adopt the mashup style application development.

References

1. HousingMaps, http://www.housingmaps.com
2. Craigslist, http://www.craigslist.com
3. Google Maps, http://maps.google.com
4. ProgrammableWeb, http://www.programmableweb.com
5. Yahoo Widgets, http://widgets.yahoo.com
6. iGoogle, http://igoogle.com
7. QEDWiki, http://services.alphaworks.ibm.com/qedwiki/

8. How businesses are using Web 2.0 - A McKinsey Global Survey,
   `http://www.mckinseyquarterly.com/Information_Technology/Appl`
   `icaions/How_businesses_are_using_Web_20_A_McKinsey_Global_Su`
   `rvey_1913`
9. Bradley, A.: Key Issues for Enterprise 'Mashup' Practices, Technologies and Products, 2008. Gartner (2008)
10. Introducing Ajax and OpenAjax,
    `http://www.openajax.org/whitepapers/IntroducingAjaxandOpenAj`
    `ax.php`
11. Next-Generation Applications Using Ajax and OpenAjax, `http://www.openajax.org/`
    `whitepapers/Next-GenerationApplicationsUsingAjaxandOpenAjax.php`
12. IBM Widgets Proposal,
    `http://www.openajax.org/member/wiki/IBM_Widgets_proposal`
13. OpenAjax Hub 1.0 Specification,
    `http://www.openajax.org/member/wiki/OpenAjax_Hub_1.0_Specifi`
    `cation`
14. Fielding, R.T.: Architectural Styles and the Design of Network-based Software Architectures. PhD Thesis, University of California, Irvine (2000)
15. The 10 top challenges facing enterprise Mashups,
    `http://blogs.zdnet.com/Hinchcliffe/?p=141`
16. Mashups - The new breed of Web app Mashups,
    `http://www.ibm.com/developerworks/xml/library/x-mashups.html`

Bill Organiser Portal: A Case Study on End-User Composition

Agnes Ro, Lily Shu-Yi Xia, Hye-Young Paik,
and Chea Hyon Chon

CSE, UNSW, Sydney, Australia
{agnesr,lilyx,hpaik,cheac}@cse.unsw.edu.au

Abstract. Whilst Web services can be composed by technical developers using a language such as BPEL, there is no easy way for non-technical end users to take advantage of these services. The advent of Web 2.0 and mashups has brought about the notion that content from different sources can be brought together by the user themselves to create a new service. Inspired by such ideas, we propose a lightweight end-user service composition paradigm, namely; *Stones, Stories* and *StoryBoard*. A Stone is a representation of a commonly performed task or operation that can be used to construct a Story. *StoryBoard* provides an intuitive drag-and-drop style user environment in which the Stories are created, validated and run. We demonstrate the concept through an implementation of a case study on bill management.

1   Introduction

The number of E-Commerce systems has grown dramatically over the last few years and now becoming a fundamental part of many businesses. However, with the large amount of information and services available on the Web, it is sometimes difficult for regular users[1] to piece it all together. Due to this problem, information and services portals (e.g., Yahoo, Expedia) are becoming increasingly popular.

With Web Service technology[7], it is possible to integrate business functionality into the one portal. This creates a potential service aggregation environment from which end users could greatly benefit. However, most portals stop short of providing their customers with the ability to aggregate services on-demand. They may support a simple, pre-defined workflow (e.g., book flight, rent a car, then book accommodation), or enable personalised configuration of individual services (e.g., Weather service for Sydney), but end users cannot 'compose' the service functionality on offer as a need arises.

Although service composition has been a well-discussed topic for the researchers and developers of Web services, the tools and methodologies for enabling end-user service composition have been largely ignored.

---

[1] We refer to them as end-users in this paper.

S. Hartmann et al. (Eds.): WISE 2008, LNCS 5176, pp. 152–161, 2008.
© Springer-Verlag Berlin Heidelberg 2008

Mashups enable users to aggregate and filter information from more than one source at one convenient location. Intuitively, the concepts represented in Mashups could be applied, not only to data but also to Web services. If business had their operations exposed through Web services, end users could utilise these services and engage them into a process to suit the individual's needs.

In this paper, inspired by Mashups, we propose an end-user service composition paradigm, namely; *Stones*, *Stories* and *StoryBoard*. A Stone is a representation of a commonly performed task or operation that can be used to construct a Story. *StoryBoard* provides an intuitive drag-and-drop style user environment in which the Stories are created, validated and run.

The paper is organised as follows: Section 2 discusses background and related work, followed by a description of the case study in Section 3. The end-user service composition framework is detailed in Section 4. Sections 5 and 6 discuss evaluation and concluding remarks, respectively.

2   Related Work

Many professional Web service developers can rely on the languages (e.g., BPEL[1]) and tools (e.g., Oracle Process Manager[11]) for service composition tasks. However, the same level of support has not been given to end users.

There has been an effort to simplify BPEL (e.g., Simple Service Composition Language[5]) or provide a guided-assistance along the composition process by modelling user preferences, past experience or service dependencies[6,13]. However, we argue that none of the approaches is intuitive to end users in that they do not hide all the complexity of underlying technology. End users still have to understand the concept behind the tools or learn a language to be able to compose even a simple process.

Mashups[8,9,3] opened up easy access to data silos which were previously only accessible through a Web site. Major IT companies now provide so-called "end user oriented" mashup tools (e.g., Yahoo! Pipes, Microsoft Popfly) enabling the users to compose and organise existing information sources.

However, current Mashup tools and their applications are focused on accessing, manipulating data and composing data flows (e.g., filtering, merging, sorting data feeds). To use the tools effectively, the users need to know, not only how to 'program', but also how to use the different Web APIs from all services[4].

Our project is inspired by Mashup techniques in that we would like to empower the end-users with intuitive tools that allow them to create 'service' from existing 'services' as freely as their personal needs dictate, and also facilitate software reuse in mass by sharing such services with others. The core idea behind the Stones and Stories in this paper is to provide an intuitive and lightweight Web service composition environment for the users to define and execute repetitive tasks by "mashing-up" available Web service operations, without realising the complexity behind.

We would like to note that a technique commonly known as 'Web Scripting' seeks to achieve similar goals[14,2]. For example, Koala[14] is a system that records the sequence of user actions during a Web browsing session. The sequence can be automatically repeated (i.e., playback) later in the future. Koala's main aim is for sharing commonly performed business process with could be shared with co-workers. However, Web scripting strictly applies to Web page browsing activities (e.g., requesting for a particular URL, filling in forms in the input boxes), not Web service operations.

3   Case Study: Bill Management Issues

To demonstrate the lightweight end-user composition concept, we take bill payment management as a case study.

The average consumer household has bills for water, electricity, telephone and gas. Some have additional bills for Internet, pay TV, health insurance, and car insurance. If the consumer owns their own home they receive bills for council rates and strata levies, otherwise they make rental payments. Individuals in the household receive bills for mobile phones, newspaper subscriptions, magazine subscriptions, and education/tuition fees. The bills may arrive monthly, quarterly, annually, and may have payment options via Post BillPay, direct debt, BPAY, Bill Express, or credit card.

Managing these bill payments is undeniably a substantial task which takes up a considerable amount of an individual's time and effort. Bills must be paid on time using an accepted payment format and paper bills need to be sorted and stored appropriately after payment. Each payment option requires the consumer to supply customer reference numbers and bill reference numbers which may change dynamically with every bill or may remain the same, depending the issuer of the bill or the payment option. It is clear that the repetitive tasks involved in the process of bill payment should be handled by a bill management system.

We implement a system named Billing Organiser Portal (BOP). In BOP, for example, a consumer may create a custom *Story* which retrieves all outstanding bills, pay them with a particular credit card and receive the receipt in an email.

Although currently there are existing systems that assist in bill management, there is currently no solution to integrate the entire bill management process from the biller to a consumer's financial service account, not to mention the ability to compose new functionality from existing services. In the following, we describe our experience with designing and building BOP.

3.1   Overview of Bill Organiser Portal

BOP uses Web services to aggregate business functionality such as the issuing and paying of bills for all bill providers as well as the displaying and managing of funds for all financial accounts at a single location customised for the individual consumer. In addition, BOP uses this portal as an avenue to explore the use of a simple, intuitive user interface for end-user Web service composition.

BOP acts as a single contact point for all businesses to interact with a specific consumer. Consumers who log on to BOP will be able to see bills issued by all of their registered billers as well as the funds available in all of their registered financial accounts. The consumer can then directly make payments for their bills from their chosen financial accounts.

Due to the repetitive nature of these bill management tasks, they are perfect candidates for Web service composition by the end user. Once a consumer has composed their own personal Web service process as *Stores*, they can run the process in one simple step, schedule the process to run at a certain point in time, or even share the process with other consumers who may wish to perform similar tasks.

3.2   Implementing the Web Application Module

The Web application module of BOP has been designed based upon the Model-View-Controller design pattern, standard for most Web applications today. The module is responsible for managing users and their billing and financial service accounts. Various Java-based application frameworks such as Hibernate (for object/relational persistence and query service), Spring(for managing dependency injection among Java objects) and WebWork (for effectively managing User Interface issues) were utilised. A typical look and feel of the application is shown in Figure 1. Full implementation details can be found in [12].

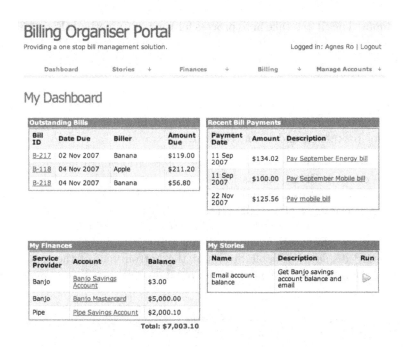

Fig. 1. A typical user interface: BOP Dashboard Screen

3.3   Implementing the Web Services Module

For implementation purposes, we have defined two separate interfaces for Billing and Financial Service. External businesses that interact with BOP are expected to adopt the interfaces. It would be ideal if the interfaces for such operations were defined by an industry acknowledged Web services standards body. However the generalisation difficulties in standards definition for such industry specific functionality mean that such standards are not in existence[10].

The following snippet shows the interface definition that Billing Service providers must implement in order for BOP to interact directly with their system. Evidently, billing providers must provide the critical functionality, for example, to view individual consumer bills online (e.g., `getOutstandingBills()`). Similarly, there is also an interface required to be implemented by Financial Service providers.

```
public interface BopBillingService {
  public String authenticate(); // Returns xhtml form to be displayed for authentication
  public Bill[] getOutstandingBills(); // Returns a list of Bill Objects
  public String[] getPaymentMethods(); // Returns a list of Payment methods accepted by the biller
  public String getBillerCode(); // Return the BPAY Biller code of the biller
  public AccountDetails getAccountDetails(); // Return the biller's account details
  public String payBillViaCC(); // Pay a bill with the given credit card details
}
```

4   End-User Composition in BOP

Stones and Stories: A Stone is a representation of a commonly performed task or operation that can be used to construct a Story. Each Stone has a set of defined input types it can take in to process for the `run()` method (Figure 3). The output type for each Stone is the output type returned by the `run()` method. For our case study implementation, we defined eight implementations of the Stone interface.

A Story is a sequence of Stones. In order for a Story to be valid, the output type of a Stone must be an input type for the following Stone. For example, the `GetBillsStone` returns a `BILLS` type and the `PayBPAYStone` accepts `BILLS` as its input for processing and so on (Figure 2).

StoryBoard: StoryBoard is the end-user composition environment in which Stories are created. Giving the user the power to compose Web service composition in a simple and user friendly interface, proved to be a difficult task. The final design is a product of several cycles of usability testing. We concluded that a drag/drop approach using Javascript, where Stones could be 'dragged' and 'dropped' onto a *Storyboard* would be the most intuitive design.

StoryBoard displays available Stones at the bottom-half of the screen. The top-half of the screen is the composition area into which the Stones are placed.

Fig. 2. Connecting Stones through Input and Output

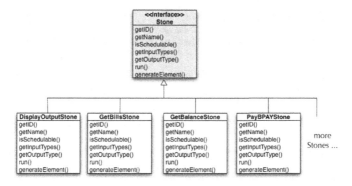

Fig. 3. The Design of Stones

StoryBoard also adds a simple validation support for the user. To ensure a valid composition is constructed by the user, only the Stones that can be placed on the next step in the Storyboard are highlighted and enabled (i.e., draggable). The other Stones are grayed out and disabled as illustrated in Figure 4. Also, the user is not allowed to skip a step (i.e., leave an empty step in between Stones). When a Stone requires inputs from the user, a popup window is automatically displayed when the Stone "clicks" into its place (Figure 5).

The actual content of the Story stored is a string of XML, containing the sequence of Stones to be executed as well as the parameters for each Stone. The

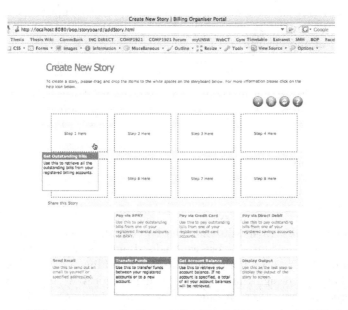

Fig. 4. For usability purposes, a toolbar at the top right corner was added for saving the Story, loading a shared Story, refreshing the storyboard and Story help. This figure shows 'Create New Story' screen.

Fig. 5. A Stone automatically displays input boxes for required inputs

schema definition for the Story content is not shown in the paper for space reasons. The `StoryManager` class is responsible for the processing and management of Stories. It relies on the `StoryDAO` class to store/retrieve/update Stories to the database, as well as the `StoryParser` class to parse the Story content from XML to Object form and vice-versa.

Once a Story is created, it can be made visible to other users in the system. A shared Story is treated like a template. Such a template contains a sequence of Stones, but no input parameters are associated with each Stone. When a shared Story is imported by a user, the user needs to update the parameters appropriately (e.g., credit card detail) before running the imported Story.

5   Evaluation

5.1   User Evaluation

Throughout the design of the system, we conducted usability testing with users and improved the design and functionality based on the feedback. This cycle was repeated a number of times before we finalised our design. Initially, most users seemed to have difficulty in creating a Story and adding a new billing account. The user interface design that was substantially improved via user evaluations was the create story screen. Ratings range from 2 to 5 as improvements were continually made in between evaluations.

In the final evaluation, total of 10 users were asked to perform the evaluation survey. Results are summarised in Figure 7 and 8. We asked the users to rate con-

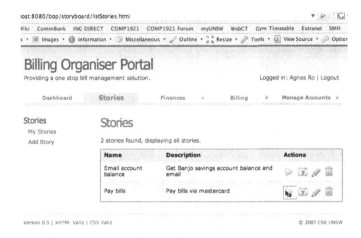

Fig. 6. The user can click on the Run icon to execute a composed Story

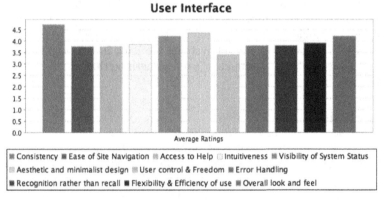

Fig. 7. User Interface Evaluation

sistency, ease of navigation, access to help, intuitiveness, visibility of system status, aesthetic and minimalist design, user control and freedom and error handling. All criteria scored minimum of 3.8/5 or higher, except for user control and freedom which rated 3.4. This indicated that the system's lack of support for undoing user mistakes. Majority of the users indicated that they would use the system if publicly available and agreed that such system will be very useful for their daily lives.

5.2   Developer Reflection

The current implementation of BOP is a full working system, that effectively demonstrates the concept of end-user composition in a business portal.

One of the major weakness in our system, is admittedly security. We are well aware that the information passed throughout the system and to the service providers is highly critical and a target for abuse. For this case study, we have only applied minimal security measures such as encryption of passwords.

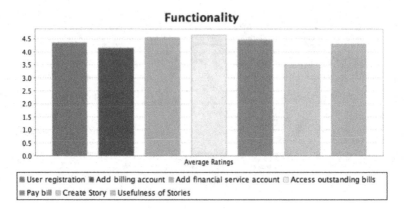

Fig. 8. Functionality Evaluation

With the integration of multiple business services in BOP, it is difficult to maintain the ideal ACID properties for transaction management. For a system that primarily deals with finances being credited/debited between billers and financial service providers, it is vital that in the failure of such a transaction, the entire process is rolled back and all systems are back in it's original state. Although this issue has been identified, it is not something that our implementation currently supports.

6   Conclusion

Web services and the advance of SOA have allowed businesses to redesign their internal systems into modularised services that they can expose to other services in their domain as well as to the world. Whilst these Web services can be composed by technical developers using a Web service composition language such as BPEL, there is no easy way for non-technical end users to take advantage of these services. This project is a study of the practicability of a simple, intuitive user interface allows end users to create their own Web service composition using services from various different sources together in a manner which is appropriate for their own personal needs.

The development of the Story/Stone Web service composition module of this paper proposes an abstraction of Web services from different sources which allows the end user to compose their own composite services without knowing about the low level details of such an endeavour.

References

1. BPEL. Business Process Execution Language for Web Services,
   http://www.ibm.com/developerworks/library/specification/ws-bpel
2. Hupp, D., Miller, R.: Smart bookmarks: Automatic Retroactive Macro Recording on the Web. In: Proc. of ACM symposium on User Interface Software and Technology, pp. 81–90 (2007)

3. Huynh, D., Karger, D., Miller, R.: Exhibit: Lightweight Structured Data Publishing. In: Proc. of IntConf on World Wide Web, pp. 737–746 (2007)

4. Di Lorenzo, G., Hacid, H., Paik, H., Benatallah, B.: Mashups for Data Integration: An Analysis, School of Computer Science and Engineering, University of New South Wales, Technical Report 0810 (2008), http://cgi.cse.unsw.edu.au/~reports/

5. Gavran, I., Milanovic, A., Srbljic, S.: In: Proc. of 7th Workshop on Distributed Data and Structures, Santa Clara, CA (2006)

6. Han, J., Han, Y., Jin, Y., Wang, J., Yu, J.: Personalized active service spaces for end-user service composition. In: Proc of IEEE International Conference on Services Computing (SCC 2006)

7. Huhns, M., Singh, M.: Service-Oriented Computing: Key Concepts and Principles. Internet Computing 9(1), 75–81 (2005)

8. Maximilien, M., Wilkison, H., Desai, N., Tai, S.: A Domain Specific Language for Web APIs and Services Mashups. In: Proc. of IntConf on Service Oriented Computing, pp. 13–26 (2007)

9. Diaz, O., Perez, S., Paz, I.: Providing Personalized Mashups Within the Context of Existing Web Applications. In: Proc. of IntConf on Web Information Systems Eng., pp. 493–502 (2007)

10. Zimmermann, O., Milinski, S., Craes, M., Oellermann, F.: Second Generation Web Services-Oriented Architecture in Production in the Finance Industry. In: Proc. of IntConf on Object Oriented Programming Systems Languages and Applications, pp. 283–289. ACM Press, New York (2004)

11. Oracle. Oracle BPEL Process Manager, www.oracle.com/appserver/bpel_home.html

12. Agnes, R., Xia, L.: End User Web Service Composition for a C2B Portal, School of Computer Science and Engineering, University of New South Wales, Thesis http://www.cse.unsw.edu.au/~hpaik/pdf/3100983.ThesisB.pdf

13. Bova, R., Paik, H., Hassas, S., Benbernou, S., Benatallah, B.: On embedding task memory in services composition frameworks. In: Proc of 7th International Conference on Web Engineering, Como, Italy, July 16-20, pp. 1–16. Springer, Heidelberg

14. Lau, T.: Social Scripting for the Web. Computer 40(6), 96–98 (2007)

Extending Services Delivery with Lightweight Composition

Christian Janiesch, Kathrin Fleischmann, and Alexander Dreiling

SAP Research Brisbane, SAP Australia Pty Ltd,
Level 12, 133 Mary St, Brisbane QLD 4000, Australia
{c.janiesch,kathrin.fleischmann,alexander.dreiling}@sap.com

Abstract. The tertiary sector is an important employer and its growth is well above average. The Texo project's aim is to support this development by making services tradable. The composition of new or value-added services is a cornerstone of the proposed architecture. It is, however, intended to cater for build-time. Yet, at run-time unforseen exceptions may occur and user's requirements may change. Varying circumstances require immediate sensemaking of the situation's context and call for prompt extensions of existing services. Lightweight composition technology provided by the RoofTop project enables domain experts to create simple widget-like applications, also termed enterprise mashups, without extensive methodological skills. In this way RoofTop can assist and extend the idea of service delivery through the Texo platform and is a further step towards a next generation internet of services.

Keywords: Service delivery platform, enterprise mashup, lightweight composition, internet of services, service ecosystems.

1 Introduction

The tertiary sector is the biggest employer in most developed countries. Its growth is above par. Not only individuals provide services but also service networks of independent companies [9]. Services are the key part of future business value networks. They are estimated to have the biggest share of the added value in the future. In order to leverage this potential, services have to become tradable goods – similar to manufactured goods.

To further support this development, an infrastructure is necessary which can provide services over the internet. The composition or aggregation of different services is a cornerstone functionality to enable the development of new, innovative services on the basis of existing services. Services can be offered and integrated by different parties. The focus of Texo [16] are web-based services, which are accessible over the internet (so called e-services). The goal of the Texo project thus is to conceptualise and implement a comprehensive solution comprising platform, models, methods, and components to support and realize dynamic business networks. All of these technologies are employed to design services, i.e. mainly to support their evolvement at build-time.

However, at run-time exceptions may occur which require immediate sensemaking of the situation's context. Furthermore, user expectations may change and cannot be

S. Hartmann et al. (Eds.): WISE 2008, LNCS 5176, pp. 162–171, 2008.
© Springer-Verlag Berlin Heidelberg 2008

catered for anymore at runtime. Decisions may have to be made that call for prompt extensions of existing services. Thereby, timely accurate and relevant information is a key contributor to making good decisions. It is necessary to put information into a certain context specific to the situation. Combining information from different sources can help answering different questions such as: What? (e.g. stock gains or loss), why? (e.g. the CEO resigned) to whom? (e.g. contextualization with profile services). The provision of this kind of information needs to be realized in an ad-hoc manner: Pre-defining certain patterns or scenarios with respect to sensemaking proves difficult since critical situations mostly arise out of unexpected circumstances. Hence, a system to support business users in such situations must enable them to identify relevant information blocks and contextualize them on the fly. Business users possess domain knowledge but not necessarily the method knowledge or time to generate a solution from scratch. They need be able to interact with the system in a non-technical way with an appealing user interface experience and an intuitive workbench. RoofTop provides such a technology [15].

This paper provides a use case and first demonstrator of ideas on the meaningful integration of enterprise mashups into next generation service provisioning in the internet of services. The paper is structured as follows: First, the architectures of both underlying technologies, Texo and RoofTop, are introduced. Second, a scenario is discussed, which points out the benefits of a joint application of both technologies. The paper closes with a summary of related work and concluding remarks on further research.

2 Texo and RoofTop Architectures

2.1 The Texo Service Delivery Platform

The focus of Texo is on the supply of tradeable business services for customers whose software is capable of using the e-services of the (Texo) platform. One of the main applications of the platform is the consumption of value-added services or additional (special) services in order to support changing requirements of the market and/ or customers. Small and medium sized enterprises are offered the opportunity to make their business service(s) available to a large number of customers without the need to have their own sales organisation in place. Individuals or organisations on the demand side can use the services directly without having to customize them for their specific needs by going through a cumbersome configuration process.

Texo is based on the principle of service orientation, which aims at offering functionality in small, loosely coupled building blocks (services) [6]. By combining these blocks, greater flexibility is achieved, allowing companies to react quickly and with agility to changing requirements while at the same time reducing cost for the services' users. For this purpose, all big players such as IBM, Microsoft and SAP have started to adjust their systems to service orientation based on open standards, e.g. SOAP and WSDL. SAP for instance announced to use their version of Service Oriented Architecture (SOA), called Enterprise SOA, as the basis for any new products [14]. The focus clearly is the offering and distribution of e-services. The general stakeholders in the architecture are service consumer, service provider, and service broker (cf. e.g. [2, 5, 7]). As mentioned above, *service consumers* can be individuals, companies or even public administrations. The consumers communicate through a so-called service delivery platform managed by a *service broker*. *Service providers* register their e-service portfolio on the platform in order to be brokered.

2.2 RoofTop

RoofTop is an AJAX-based Web 2.0 Application that allows the user to systematically contextualize and combine unstructured information from the web with structured business content from an enterprise system. It is an easy to use lightweight composition platform that enables a user to create information mashups without the need to run formal, time consuming software creation processes. The application is targeted at business users and designed based upon Web Services [17], Enterprise SOA [14] and open web standards such as RSS feeds.

The RoofTop platform offers the users multiple frames to work in: On the left hand side the user finds a list of available services from a RoofTop services repository. These services represent building blocks of various sources of information such as enterprise services or unstructured data from the web. Using the service repository, the user can connect data sources to RoofTop, i.e. create services on the RoofTop platform or update them. Every service is – depending on its service type – described by a set of parameters defining the way it connects to a source of information and how it is visually rendered (e.g. table, pager, map etc.). For existing service types (e.g. RSS), the creation of new services in the RoofTop environment does not involve technical tasks such as the deployment of new software components. Adding services is more of a customization task that can be done by technology savvy business users. New standards such as RSS or ATOM leverage the full potential of the application enabling to connect to large repositories of information on the internet. As for enterprise content, generic REST based service interfaces have proven to be a viable option for enabling the use of enterprise services on the RoofTop platform.

By dragging a service from the repository onto the work area, a running instance of the service is immediately initialized with live data. The service instances allow for further customization. For example, if the data provided by a service is rendered as a table, it is possible to enable or disable columns and limit the number of displayed rows. Runtime and design time of the system are highly intertwined: In the design time view, user interactions have an immediate effect on the data display, for instance columns can be switched on and off on the fly without requiring an additional page reload. Furthermore, all data displayed in design time is live data. In turn, the runtime simply provides a snapshot of the current configuration of a mashup page. In order to build mashups, the data sources of the running service instances need to be synchronized. For example, the contextualization of stock exchange data with Google News is done as follows: By drawing a line between the company name of the stock exchange service and the search input for the Google News service, the company name is passed on to the Google service as a search parameter. By clicking on the different stock symbols, the content of the news service is updated with news for the respective company. Figure 1 provides a screenshot of the application.

With a multitude of input parameters and potentially large sets of return data, the configuration of services may turn out to be a rather complex for a business user. This is particularly the case for enterprise services. In order to facilitate the creation ofsituational mashups involving more complex services, an assisted mode was introduced

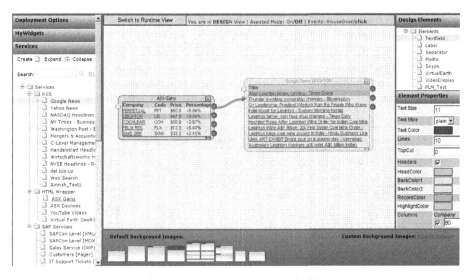

Fig. 1. Screenshot of the RoofTop Workbench

Fig. 2. Assisted mode for mashup creation

in RoofTop. In the assisted mode, services are pre-wired, i.e. it is specified which service has which predecessor and how the services are connected. By enabling the assisted mode, each service on the work area is enhanced with a drop down list of preconfigured follow up services (cf. Figure 2). When clicking on a follow up service from the list, the corresponding service opens in the work area and is automatically connected to its predecessor in a suitable way.

As a prerequisite for the assisted mode, services have to be pre-defined with respect to their inputs and outputs. At present the RoofTop prototype only offers manual pre-wiring of services. The extent of automation of this task by applying semantic technologies is subject to further research.

3 Application of Lightweight Composition to Extend a Service Delivery Case Study

3.1 Eco Calculator Scenario

In the following application scenario, two stakeholders are engaged in a transaction using the Texo marketplace. The first stakeholder is a manufacturer of car seats. This company needs to have their product checked and certified under ecological aspects with regards to the environmental sustainability of materials that are being used in car seats. The certification is needed due to new export guidelines in Australia, which is an important export market for them. The second stakeholder is a service provider who is offering a tradeable service to calculate eco values for products including certification. The Texo marketplace enables both parties to dynamically engage in a business relationship with one another. Figure 3 shows the underlying process.

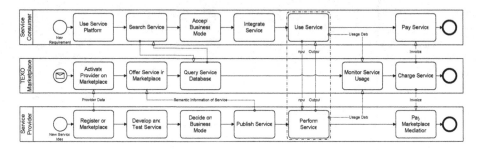

Fig. 3. General process of using the Texo marketplace

As Texo is used to sell services (e-services to be more specific) through a marketplace, all service providers are required to register with this intermediary. Depending on the type of service offered, it might also be necessary that the supplier is able to provide a certification for certain systems. This way, seamless integration of the offered service with the customer's system can be guaranteed. A number of business models are possible in this context: e.g. regular charges or usage based fees. The prerequisite for the customer is accordingly. He has to use software which is able to communicate with the Texo marketplace and the services offered through it. This basic requirement is essential in order to ensure a smooth integration of the e-service into the user-interface of the customer's system. This is the precondition for a dynamic integration and composition of services.

As the company wants to export car seats for a car manufacturer to Australia, it needs a certificate which is currently not offered by its own system. Therefore, it queries the Texo marketplace using text including keywords describing the requirements to obtain the certificate. As a result of the query, services are suggested which match the keywords and the inquiring company can choose the service which best matches their needs. Using the Texo platform as a basis, the service can be used without having to run a manual integration project.

Figure 4 shows the subsequent integration of the service into the user interface of the inquiring company. The compact display for the developer is extended by an additional column on the right side (eco value), the value of which can be calculated taking the type of materials used in constructing the car seat into account. The example shows an achieved total value of 85 %, which is sufficient to be certified. The inquiring company can therefore request the certificate for their car seat from the service provider and save it in their system. During this process, Texo is running as a background service and is collecting usage statistics that can be used for charging as well as potentially for service improvement and as a general feedback mechanism.

Fig. 4. Service usage through the Texo platform

3.2 Service Extensions with Lightweight Composition

If, however, in the above scenario the total value of the eco value was below 85 %, a situation arises, which requires flexible problem solving capabilities and on-the-fly sensemaking to evaluate alternatives. RoofTop provides this instant lightweight composition of enterprise mashups. This enables the service user to extend the functionality and/ or use related technology to solve an acute problem. If we assume the value returned by the eco calculator turns out to be 80 %, the manufacturer might now be interested in evaluating the cause of the problem. That is, he would like to know due to which component the value is as low as it is. Furthermore, he is interested in taking counter measures such as changing a certain material in the production process of the car seat – if this is possible at all given that changing a material may incur further consequences such as switching to a different supplier. In the present case study the information about the used production model components of the car seat is the starting point for further sensemaking.

On the left hand side, Figure 5 shows a situation in the Texo enabled development environment after the calculated eco value only turned out to be 80 %. The list of components, which the car seat consists of, serves as a starting point for further investigation using lightweight composition. By clicking on the associated button the user is provided with a platform that enables further contextualization of information and helps developing an immediate solution to the problem. Please refer to the right hand side of Figure 5 for the subsequent procedure using RoofTop.

In the current situation we assume the user chooses to find out more about the materials which are used to assemble to production model components. By examining the components and the assigned materials, the user can discover that the synthetic foam material used in the car seat's component has a particular low eco value. The materials service is already part of the RoofTop service repository and has pre-configured connections from the production model components service. That means the user can simply click on the associated button in the Texo enabled development

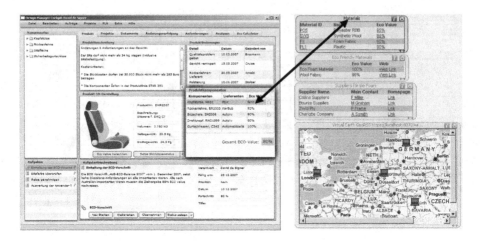

Fig. 5. Immediate service exception sensemaking and solution development

environment. An external service is then used to assess whether there are more environmentally friendly materials that could be used as a substitute for the synthetic foam material. The service was part of the eco calculator package and is automatically added to the RoofTop service repository when connecting to the eco calculator through the Texo marketplace.

In order to switch to a different material, it has to be evaluated as to whether there are suppliers for the material and as to whether the conditions of the prospective supplier are suitable. Using an existing enterprise service as part of the RoofTop service repository the user can look for suppliers for the eco-friendly materials. Furthermore, it is of interest where the suppliers are located. The mashup environment provides a map rendering option for plotting the locations of the prospective suppliers. Displaying the locations of the prospective supplier can help the user determine the feasibility of engaging them.

In this scenario lightweight composition technology provided by the RoofTop platform enables domain experts to create simple widget-like applications without extensive methodological knowledge. Business users are able to obtain all the necessary information within a short period of time by using integrated mashup capabilities to navigate through internal as well as external information sources. They can now make an informed decision as to which material could be switched to and choose a particular supplier to optimize the bill of material for the car seat.

4   Related Work

Services can be defined as not storable and intangible goods which are constructed in cooperation with an external factor (usually the service consumer). Construction and consumption traditionally occur at the same time (uno-actu principle). Electronic services (e-services) differ insofar as they are storable in a sense and their consumption, i.e. execution, does not necessarily involve concurrence (for on overview cf.

[1]). Innovative Services are the successful exploitation of new service ideas and enable change which creates a new dimension of performance.

The underlying technology of a service delivery platform clearly stems from the area of service oriented architectures (SOA) [11-13] and software as a service (SaaS) [4]. SOAs provide coarse-grained loosely coupled, self-contained services which constitute the basis of enterprise end-to-end processes. SaaS is a concept to provide software from the vendor to the consumer through remote access including, e.g., maintenance and upgrades.

Nowadays several companies offer or are about to offer software, data or infrastructures as a service through the internet based on the principles of an SOA. Examples include but are not limited to Google Apps, SalesForce, AppExchange, StrikeIron Marketplace, Iceberg on demand, Amazon Web Services, or SAP Business byDesign.[1]

In the context of Web 2.0, mashup platforms such as RoofTop are gaining increasing popularity both in the consumer space as well as in the enterprise world. Gartner identified mashups and composite applications as one of the Top 10 strategic technologies for 2008 [8]. Large as well as small software vendors have started to offer mashup platforms. Examples include Microsoft's Popfly, IBM's QEDWiki, Yahoo Pipes, Intel MashMaker, JackBe among others.[2] The key differentiator of RoofTop is the ability to connect to enterprise content whilst offering an intuitive workbench targeted at business users. Apart from the tools mentioned here, significant related work can be found in the area of information retrieval: A distinction can be made between tools that assist users in extracting information out of an existing web page and reuse it for instance in a personalized portal [10] and systems that gather information in order to extract and integrate information from different sources into an information repository [3]. The focus of RoofTop is on the ad-hoc composition of information using mashups – the information extraction itself relies on standards such as RSS or ATOM, persisting data is due to its ad-hoc nature not in scope.

5 Conclusions and Outlook on Future Work

Composing and delivering services through an open platform is a major aim of the Texo project. It is, however, intended to cater for build-time. Yet, at run-time exceptions may occur which require immediate sensemaking of the situation's context or call for prompt extensions of existing services. In order to do so, various and possibly extensive sources of information need to be queried. Finding a way through vast amounts of information has been a challenge since the introduction of information technology in organizations. This problem has been fuelled with the growth of large organization-internal and -external information repositories over time.

Whereas it was rather difficult in the past to link different systems and have them interoperating according to a given business scenario, lightweight composition

---

[1] Cf. www.google.com/a, www.salesforce.com/appexchange, www.strikeiron.com/StrikeIron Services.aspx, icebergondemand.com, aws.amazon.com, www.sap.com/solutions/sme/businessbydesign/index.epx

[2] Cf. www.popfly.com, services.alphaworks.ibm.com/qedwiki, pipes.yahoo.com, mashmaker.intel.com, www.jackbe.com

effectively overcomes some of the restrictions of the past. New standards such as RSS or ATOM allow for implementing generic APIs to tap into large repositories on the Internet. In addition, the REST paradigm has proven superior to SOAP for the loose coupling between different web services. RoofTop is an application that systematically leverages these new standards and enables end users to create composite applications without implementing a single line of code. Creating such a mashup is paradigmatically different in that it is a content-oriented development style. Whereas object-oriented programming, logic programming or functional programming primarily addresses technology-savvy users, content-oriented development attracts a new type of users, to which information workers belong.

Our work has two major implications for practice. First, lightweight composition or the development of mashups in a content-oriented way drives down the total cost of ownership for composite applications as it abstracts from many technical details. Technology facilitating such a content-oriented style will certainly gain in importance in the future. Second, a new user group has been enabled with the power to create composite applications. If applied intelligently, IT departments can be relieved from some of their tasks in the area of information and application provision. In addition many communication problems between business and IT can be addressed (IT delivering something different than the business demanded).

In addition, our work has implications for academia. A lot of security-related research is necessary for mashups to scale to a primary means for providing a composite application. Examples include handling cross-domain scripting appropriately, catering for distributed authorization and federated authentication, provide means for Web Service reputation etc. Subject to our future work is the integration of semantic technologies in order support the user connecting and finding appropriate services.

Acknowledgements

The project was funded by means of the German Federal Ministry of Economy and Technology under the promotional reference "01MQ07012". The authors take the responsibility for the contents.

References

[1] Baida, Z., Gordijn, J., Omelayenko, B.: A Shared Service Terminology for Online Service Provisioning. In: Proc. 6th international Conference on Electronic Commerce (ICEC), pp. 1–10 (2004)

[2] Barros, A., Dumas, M.: The Rise of Web Service Ecosystems. IT Professional 8, 31–37 (2006)

[3] Baumgartner, R., Gottlob, G., Herzog, M., Slany, W.: Interactively Adding Web Service Interfaces to Existing Web Applications. In: Proc. International Symposium on Applications and the Internet (SAINT), pp. 74–80 (2004)

[4] Clark, L. M. et al.: Hype Cycle for Software as a Service, Gartner, Inc. Research Report (2006), http://sharepoint.microsoft.com/sharepoint/worldwide/cn/south/SaaS/Hype%20Cycle%20for%20Software%20as%20a%20Service.pdf

[5] Durante, A., Bell, D., Goldstein, L., Gustafson, J., Kuno, H.: A Model for the E-Service Marketplace. HP Labs 2000 Technical Reports No. HPL-2000-17, Palo Alto, CA (2000)

[6] Erl, T.: Service-oriented Architecture: Concepts, Technology, and Design. Prentice Hall, Englewood Cliffs (2005)

[7] Galambos, G.M.: Services Ecosystem. In: Keynote at 2005 IEEE International Conference on Services Computing (SCC), Orlando, FL (2005)

[8] Gartner, Inc.: Gartner Identifies the Top 10 Strategic Technologies for 2008 (2007), http://www.gartner.com/it/page.jsp?id=530109

[9] Kagermann, H., Österle, H.: Geschäftsmodelle 2010. Wie CEOs Unternehmen transformieren, 2nd edn. Frankfurter Allgemeine Buch, Frankfurt am Main (2006)

[10] Kowalkiewicz, M., Orlowska, M.E., Kaczmarek, T., Abramowicz, W.: Robust Web Content Extraction. In: Proc. 15th International Conference on World Wide Web (WWW), pp. 887–888 (2006)

[11] Krafzig, D., Banke, K., Slama, D.: Enterprise SOA: Service-oriented Architecture Best Practices. Prentice Hall, Englewood Cliffs (2004)

[12] Marks, E.A., Bell, M.: Service-Oriented Architecture (SOA): A Planning and Implementation Guide for Business and Technology. John Wiley & Sons, Hoboken (2006)

[13] Pulier, E., Taylor, H.: Understanding Enterprise SOA. Manning, Greenwich (2005)

[14] SAP AG: Enterprise Services Architecture: An Introduction. SAP Whitepaper (2004), http://www.sap.com/solutions/esa/pdf/BWP_WP_Enterprise_Servi ces_Architecture_Intro.pdf

[15] SAP INFO: Werkzeuge für Wissensarbeiter (2008), http://www.sap.info/public/DE/de/index/Category-28943c61b1e60d84b-de/0/articlesVersions-536747ea0a5fd4430

[16] Theseus Pressebüro: The Theseus Program (2007), http://theseus-programm.de/front

[17] World Wide Web Consortium: Web Services Activity (2008), http://www.w3.org/2002/ws/

Fixed-Mobile Hybrid Mashups: Applying the REST Principles to Mobile-Specific Resources

Sami Mäkeläinen and Timo Alakoski

Nokia Siemens Networks, Karaportti 2, 02610 Espoo, Finland
{sami.makelainen,timo.alakoski}@nsn.com

Abstract. Mashups have already for years been quite popular and common in the Internet. However, data sources used for creating mashups rarely include anything from the vast potential of the mobile domain and mashups either used from mobile devices or utilizing mobile-specific assets are largely absent. Many enablers, such as smartphones equipped with GPS receivers, already exist for such mashups; operators also possess a large array of useful information on their network server. The lack of mashups in this potentially fruitful area stems partly from the difficulties in accessing such data. This paper describers the application of Representational State Transfer (REST) principles to opening up new and legacy mobile-specific assets in order to enable the creation of compelling "fixed-mobile hybrid" mashups – mashups that utilize information both from existing "fixed" Internet sources as well as mobile-specific sources exposed via an open, simple API. We also present alternative architectures for the deployment of such a system and provide key lessons learned when implementing a subset of the designed API and proof-of-concept mashups.

Keywords: Mashups, REST, software architecture, mobile services, SDP, SDF.

1 Introduction and Motivation

Mobile operators are increasingly facing competition from the Internet; innovative new applications and services are being deployed in manners that often reduce mobile operators to bit-pipes, leaving them out of key value chains. One of the key challenges that operators face in introducing new services is how they can match or harness the speed and innovative power of the so-called "Internet developer community".

Services in the mobile operator domain have traditionally emerged through lengthy standardization processes and time to market for services was / is easily measured in years. Mobile devices, meanwhile, have developed to the point of being able to offer a platform for applications written by third parties, bypassing the traditional operator channels. With web-based applications, this trend will only accelerate – the services will slowly move to be used through the browser, as has happened on the "fixed" Internet.

To solve the time to market challenge, many operators are looking to Service Delivery Platforms (SDP) to help them accelerate the application creation and launch process [7]. For the most parts, however, the SDPs deployed are aimed toward the

S. Hartmann et al. (Eds.): WISE 2008, LNCS 5176, pp. 172–182, 2008.

operators own internal application development or to a closed group of partners. This is in stark contrast to the trends in the Internet, where open APIs allow basically anyone immediate, easy programmable access to many features of popular Internet services such as Amazon, Google and Yahoo! In recent years, these APIs have increasingly been implemented in a so-called RESTful manner – i.e. following the software architecture design paradigm called Representational State Transfer (REST). Despite some operators like Vodafone, Orange, BT and Telefonica moving to provide more open and simple APIs [4, 5, 10, 11], they are still mostly absent from the operator domain and, more important, largely unutilized.

If the operators provided "Internet-style", simple, open interface to a range of useful services and subscriber information, it would usher in completely new types of mashups that we call *fixed-mobile hybrid mashups*; mashups that combine mobile-specific information with existing services in the Internet. This became the motivation for an innovation project undertaken at Nokia Siemens Networks in 2007-2008 that this paper derives from. The goals of the project were to:

a) Based on the REST principles, design a generic, simple, flexible & reusable interface to main mobile-specific services and mobile operator assets that is easily accessible by 3rd party developers, whether individual or corporations

b) Include phone-based systems to allow the phone to be used not only for accessing mashups but also for acting as another information source

c) Create proof-of-concept hybrid fixed-mobile mashups.

2 RESTful Design of Exposing Mobile Operator Services

This chapter will provide a short introduction to REST, an overview of how and which mobile-specific services were exposed and how the API was designed using REST principles.

2.1 Short Introduction to REST

REST is a software design architecture originally introduced by Roy Fielding in his Doctoral Thesis in the year 2000 [1]. It is widely used in architecting Internet services, but has not yet gained popularity in the mobile operator domain [2]. Contrary to 3GPP standards, REST is not a standard, specification or a protocol and does not strictly dictate how services should be designed. Having said that; REST relies on the following principles [2, 6]:

• Application state and functionality are divided into resources
• Every resource is uniquely addressable using a universal syntax for use in hypermedia links; i.e. URIs are used as resource identifiers
• All resources share a uniform, simple interface for the transfer of state between client and resource, consisting of a constrained set of well-defined operations and a constrained set of content types
• A protocol which is client-server, stateless cacheable and layered

REST can also be described as being *"a small set of verbs operating on a rich set of nouns"*. HTTP is typically used as the protocol on top of which RESTful APIs are implemented as it's universally supported by the Internet infrastructure. Also, existing HTTP methods can be conveniently mapped to other well-known primitives as can be seen from Table 1 below:

Table 1. Mapping of HTTP methods to other well-known primitives

HTTP	GET	POST	PUT	DELETE
CRUD	Read	Create	Update	Destroy
SQL	SELECT	INSERT	UPDATE	DELETE

As opposed to many traditional Web Services-style APIs, REST-type APIs are easily understood by humans reading them. The paradigm is most clearly illustrated by an example; the following is fictitious example of a SOAP-request for a weather forecast:

```
<SOAP-ENV:Envelope xmlns:xsd="http://www.w3.org/2001/XMLSchema"
    xmlns:SOAP-ENV="http://schemas.xmlsoap.org/soap/envelope/
    xmlns:xsi="http://www.w3.org/2001/XMLSchema-instance">

<SOAP-ENV:Body>
        <ns1:getCityWeather xmlns:ns1="Weather">
        <op1 xsi:type="xsd:string">Singapore</op1>
        </ns1:getCityWeather>
</SOAP-ENV:Body>
</SOAP-ENV:Envelope>
```

Compare this to the REST-style equivalent performed with a clear, easily readable HTTP-request that is also a more compact representation:

```
GET /restapp/weather/Singapore HTTP/1.1
HOST www.yoursite.com
```

2.2 Selecting Services to Expose

To begin the process of designing a RESTful open API for mobile operator services, one must first determine which functionality and data should be exposed. To do this, a dozen service scenarios focusing on social networking, location and communications were sketched and then analyzed to see what features implementing them would require. Some of these scenarios were later chosen to be implemented as proof-of-concept services.

While the exact set of features and functions to expose in real-life deployments will naturally depend on the desires and goals of the deploying operator, the following functionality was deemed as most important to be exposed over the REST-type API. The list is based on services and features that were deemed most useful and widely used (such as SMS sending), most uniquely mobile (such as location) and most exclusively mobile operator-owned information (such as subscriber information):

- Access to subscriber status data such as presence status and location
- Subscriber profile information such as name, address, photo, account balance, billing information etc.
- Access to subscriber voice mails and call logs
- SMS and MMS messaging functionality
- Search features; e.g. searching subscribers based on location
- Statistics

Access to all of the functionality would obviously be granted only with appropriate access rights. The authorization and security model of the solution is discussed in more detail in chapter 3.6.

2.3 RESTful API Design and Examples

With the basic functionalities and information to be available via the API decided, the API itself needs to be designed. Following the abovementioned principles of REST, it needs to be kept in mind that everything need to be exposed as resources and all activities need to be performed via the few available primitives. Therefore, one of the key decisions when designing a RESTful system is to develop a clear and concise URI structure.

The chosen URI structure, which is mostly self-explanatory, was the following:

```
https://op.com/users/<UID>
https://op.com/users/<UID>/status
https://op.com/users/<UID>/sms
https://op.com/users/<UID>/balance
https://op.com/users/<UID>/location
https://op.com/users/<UID>/voicemail
https://op.com/users/<UID>/...

https://op.com/search/location/?...
https://op.com/search/users/?...

https://op.com/stats/users/...

https://op.com/info/network/...
```

As can be seen, a structure based on the user ID (denoted by <UID> above) was chosen. The reasoning behind this decision is that most data usage will take place on individual users or on a small set of users. Another main structure is related to the search functions, with the /search/ URI "header". In the search URIs, search terms come after the trailing question mark and can include various parameters based on the information being searched. For most URIs, both read and write-operations are supported providing the requestor has sufficient access rights to perform the operation.

Third, various statistics can be made available under a URI structure preceded by /stats/. Finally, information about network elements (such as the geographical location of the base stations) can feasibly be fetched under the /info/ structure. Detailed structure of these is omitted from this paper due to space limitations.

As an example, to fetch the location of the user (in the form of ICBM location) with the user id "jdoe", the following request is made:

```
GET /users/jdoe/location HTTP/1.1
Host: op.com
```

The reply will come in the response to the HTTP request. Similarly, to update the location of the same user – provided that the entity making the request has sufficient privileges – the following request is made. This request can be used, for example, when a user wants to self-update his/her location from the device using e.g. the GPS receiver on the device, thus enabling better positioning accuracy than would be available purely from the operator's network.

```
PUT /users/jdoe/location HTTP/1.1
Host: op.com

Location: 60.1685;24.9425
Accuracy: 20m
```

2.4 Notes on Related Work

Mobile mashups are not, as such, a novel or a new thing. For example, platform frameworks to facilitate mashup creation on high-end mobile devices have been presented [8]. The participation of mobile devices in mashup creation and/or access is generally seen as potentially useful. Specifically, the location information available from GPS-equipped devices is seen as an important element in mashup creation [8, 9]. However, to our knowledge there have been few or no previous attempts at creating a uniform, simple, open interface for acquiring information such as the mobile device's location whatever the underlying positioning technology may be.

On the operator side, a few major operators have already launched APIs to open certain functionality to registered developers. In the first stages, most APIs offer relatively simple features like sending of SMS messages and other basic functionality [4, 10, 11]. However, while there have been no plans to include mobile-originated data over the APIs, the trend is clearly towards increasing interest in such APIs and operators also plan to open more complex features such as billing interfaces in the future [11].

3 Solution Architecture

The RESTful API itself is only the outward-looking interface of the system; how the back-end is implemented is a different topic that is discussed in this chapter. First, the overall actors in the ecosystem are covered. Then the solution architecture options are discussed and finally the security architecture is presented.

3.1 Actors in the Ecosystem

A simple API itself doesn't solve anything – the RESTful API to the mobile operator services is just one piece of the puzzle. There are many other, equally important pieces in the complete "mashup-ecosystem". The actors present and their interrelationships are:

- **User** – end users using the mashups via their browser, either on a PC or a mobile device
- **Mobile device** – typically a smartphone but can also be an Internet tablet-type of a device. They can also be used as sources of information transmitted to the RESTful API system.
- **Mobile operator** – typically manages and provides the RESTful API to the mashup developers for use in mashup creation.
- **Internet service provider** – an existing 3rd party providing some service via an open API for developers to use. For example, Google, Amazon, Flickr etc. They may or may not have a relationship with the operators or the mashup developers who are using their APIs.
- **Mashup developer** – typically an external 3rd party, either an individual or a corporation, developing mashups using the RESTful API and other relevant APIs to implement the required use cases. Developers may or may not have a relationship with the parties whose APIs they are using.

3.2 Optional Device Component

In addition to server-side elements, there can be an optional mobile device component present. This enables the delivery of accurate location and other information directly from a capable mobile phone to the RESTful API server.

In our proof-of-concept implementation, this was done by using a Python-application on Nokia N95 terminals. In addition to GPS location information, it collected data on presence status, active phone profile and ring tone, calendar status (busy/free), signal strength and battery level. The phone client periodically sends a status update with all available information to the server.

3.3 Option 1: Standalone System

Before handling the architecture options, it is important to point out that the RESTful API can be considered as not just an API but rather a lightweight, fully functioning self-contained system capable of providing a limited set of services via an open API even without any back-end support. The architecturally simplest solution is therefore to run the RESTful API and the required server-side functionality as a standalone system. In this scenario, the system can be run by anyone as no integration to operator systems is required. The simplified architecture diagram of the standalone system is presented below in Figure 1:

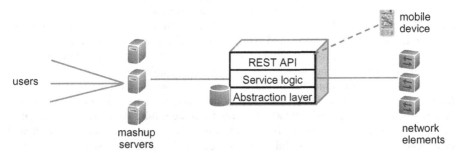

Fig. 1. Standalone system architecture

There are, however, also significant disadvantages in using a standalone system. While fetching the subscribers' location and a range of other information is possible using the device component covered in Chapter 3.2, most subscriber information resides on operator systems and would be inaccessible in this scenario. Additionally, subscriber autoprovisioning is impossible if pre-existing subscriber records are non-existent and the device-component will somehow need to be installed on the phone.

For these and various other reasons, the standalone system is deemed feasible really only for limited deployments, e.g. for testing and development purposes.

3.4 Option 2: Integration Via SDP-Style Middleware

In order for the operators to gain the benefits of the RESTful API, they will naturally need to integrate the API elements to the "real" back-end services such as the subscriber database, location information server etc. The SDP discussed earlier provides one method of tapping to the operator infrastructure without direct integration to all the required network elements.

In essence, the RESTful API can be implemented as an application either inside or outside of the SDP, providing another kind of interface to the outside developers and letting the SDP platform worry about the internal interface details. This architecture is presented in Figure 2 below.

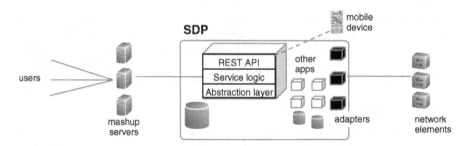

Fig. 2. Integration with an SDP-style middleware

3.5 Option 3: Integration Directly to Network Elements

Another option of integrating the RESTful API into the operator systems is to have the RESTful API itself reside on a system which is integrated directly, on an as-needed-basis, to elements in the operator infrastructure. This approach is most feasible for operators that do not intend to deploy an SDP as this route allows them to bypass the unnecessarily heavy middleware that the SDP would become if only deploying the RESTful API is the goal. This architecture is presented in Figure 3.

Yet another option, closely related to integration directly with the network elements, is for the network elements themselves to provide a simple, open API. However, this approach suffers from a number of drawbacks; first, it does not offer a unified view to the information available. Second, security policies and network configuration typically prevent the direct access of network elements from the public Internet, thus necessitating a proxy or a similar central element anyway. Finally, in a typical network, it will take many years for all the useful network elements to be upgraded to versions that provide an open API, thus crucially slowing down the rollout of the interface.

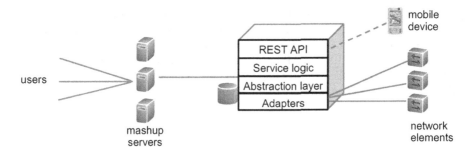

Fig. 3. Integration directly to network elements

3.6 Security Model

One of the most important aspects when opening access to any operator systems is security; it is essential to control who have access to critical information. Yet at the same time, access to the APIs should be as simple as possible and convenient for the users of the API.

Starting from these premises, it was analyzed what kind of security model was appropriate for the RESTful API. The following requirements were set for the system:

- It must be possible to authenticate and authorize all requests if so desired
- Access to subscriber information must be controlled with access control rules allowing for fine-grained control
- Subscribers must have full control on which external parties are allowed to access their information and to what extent access is allowed

After analyzing possible options, it was decided that the new OAuth [3] specification would fit the authorization purpose well. User ID & passwords transmitted over an encrypted link are used for authentication. OAuth solves the problem how the mashup developer (Service Provider) can access the user's information from the operator without giving his or hers credentials to the developer.

Our OAuth implementation was enhanced so that it is possible for all users to grant or deny access per resource requested – this means that user A can grant access to Service Provider X to access his or her location information but deny access to his / her profile data, even if the X had originally requested full access to A's information. Users can also do very fine-grained access control to resource, e.g. grant Service Provider X access only to the user's resources for 1 day or for 10 times.

3.7 Proof-of-Concept Mashups

We created a standalone system to demonstrate the use of the RESTful API and the OAuth authorization mechanism. For the demonstration system, a modified version of the architecture option 1 described above was selected. The system was in essence standalone, but with pre-existing API "hooks" to an SDP, making future integration to a live network easy. An external 3rd party SMS gateway provider was utilized for sending short messages.

In our proof of concept we have two main elements; the RESTful API server and the mashup server. Lacking access to real live network elements, we created a client to the mobile phone to send the location and presence data to RESTful API server (using the specified RESTful APIs). The mashup server logic was implemented using PHP, the RESTful API with Java and the mobile client with Python for S60. Figure 4 illustrates the overall architecture of the proof of concept.

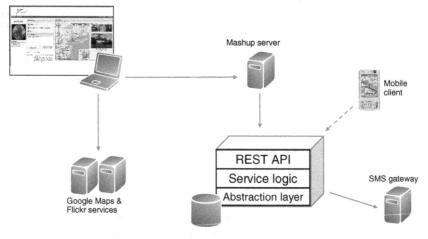

Fig. 4. Demo architecture

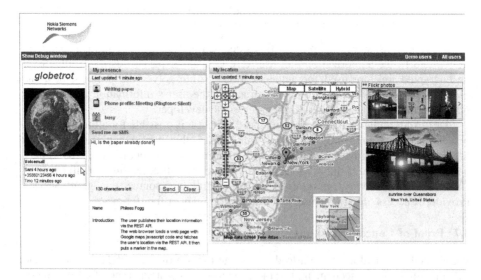

Fig. 5. Web-interface of the proof-of-concept mashup

Figure 5 shows the end-user's browser-based interface to the mashup. The user's picture and profile info is simply fetched from the server. The mashup uses the location resource of Google Maps to display the location of the user in the map. Likewise the Flickr APIs with location info are used to display the available photos from areas geographically close to the user's current location. The Presence box shows the available presence status values from the mobile client and the SMS box allows sending of SMS messages to user via RESTful API call. The voice mail box shows the available voice mails for the user and a link the actual voice mail which can be listened to via the web browser.

We also created a Facebook Flash application (see Figure 6) utilizing the same RESTful APIs to demonstrate how easy it is to create small applications and widgets and deploy them in different systems. The Facebook application uses Yahoo! Maps to show the location and presence in the user's own Facebook page.

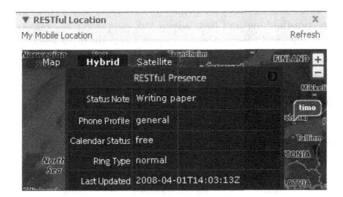

Fig. 6. Facebook application

4   Conclusions, Lessons and Recommendations

During the course of the project, it became clear that mobile-operator services and resources can be successfully exposed the "Internet-way", using simple, open APIs. In short, designing systems in a RESTful manner works even on top of mobile operator systems that have been designed using very different paradigms.

When designing an API that complies with the REST ideology, it is important to take into account some idiosyncrasies that are often present in the mobile operator domain. One of them is security; operators are traditionally very worried about security. Reconciling the need for simplicity and openness with the operators' strict security requirements is by no means an easy task but we believe utilizing OAuth-style authorization and TLS encryption for the API access provide the necessary framework for implementing a security architecture that fulfills the needs of both the Internet developers and the operators.

To operators used to dealing with complex interfaces, the promise of a simple, yet powerful API may seem dubious. Yet simple APIs do not automatically mean inflexible or somehow inferior functionality when compared with the traditional, rather complex, typically 3GPP-specified APIs prevalent in the operator domain. With careful planning, a simple and easy-to-use API can be planned and deployed that still offers remarkable flexibility and allows 3rd party Internet developers to easily deploy advanced mashups using methods and software they have become accustomed to and without mobile domain-specific special knowledge.

While from a technological point of view the findings are encouraging, the more difficult challenges lie elsewhere; the challenge in integrating the "Internet-world" with the mobile operators' domain is less about technology and more about reconciling the differences in operational and development philosophies, processes and degrees of openness. For one, mobile operators need to realize that new, innovative services are not necessarily instant financial successes, but rather provide a differentiation opportunity.

Even the best technical solution by and itself does not allow new, Internet-style innovation to magically happen in the operator domain. The new options and degrees of openness available must not only be properly implemented but also wholeheartedly adopted in the core thinking and then advertised, supported and encouraged by the operator – not immediately limited and charged. Recently there have been some encouraging developments indicating a slowly changing mindset with operators launching more open APIs and even talking about mashups. However, there is still a long road to go before fixed-mobile hybrid mashups become as commonplace services as current Internet mashups are today.

References

1. Fielding, R.: Architectural Styles and the Design of Network-based Software Architectures. Doctoral dissertation, University of California, Irvine (2000),
   http://www.ics.uci.edu/~fielding/pubs/dissertation/top.htm
2. Representational State Transfer, Wikipedia (2008),
   http://en.wikipedia.org/wiki/Representational_State_Transfer
3. Atwood, M., et al.: OAuth Core 1 specification (2007),
   http://oauth.net/core/1.0/
4. Vodafone Betavine, Vodafone (2008),http://www.vodafonebetavine.net/
5. Open movilforum (2007), http://open.movilforum.com/en/
6. Richardson, L., Ruby, S., Hansson, D.H.: RESTful Web Services. O'Reilly Media Inc., Sebastopol (2007)
7. Service Delivery Platforms, Wikipedia (2008),
   http://en.wikipedia.org/wiki/Service_Delivery_Platform
8. Brodt, A., Nicklas, D.: The TELAR Mobile Mashup Platform for Nokia Internet Tablets. In: EDBT 2008: Proceedings of the 11th international conference on Extending database technology (2008)
9. Soon, C.J., Roe, P., Tjondronegoro, D.: An approach to Mobile Collaborative Mapping. In: SAC 2008: Proceedings of the 2008 ACM symposium on Applied computing (2008)
10. Orange: Access Orange APIs (2008),
    http://www.orangepartner.com/site/enuk/access_orange_apis/p_access.jsp
11. Informa Telecoms & Media: Telecom Markets, Issue 566, May 6 (2008)

Engineering Issues for the Web 2.0

Yanchun Zhang[1], Florian Daniel[2], Santiago Melia[3],
Katsumi Tanaka[4], Athman Bouguettaya[5], and Daniela Nicklas[6]

[1] Victoria University, Australia
[2] University of Trento, Italy
[3] University of Alicante, Spain
[4] University of Kyoto, Japan
[5] CSIRO ICT Centre, Canberra, Australia
[6] Department for Computer Science,
Carl von Ossietzky Universität Oldenburg, Germany

Web 2.0 is a technical term describing the trend in the use of World Wide Web technology and web design. The aims of Web 2.0 are to enhance creativity, information sharing and collaboration among users, rather than just retrieving information. For example, one of the most promising area emerged recently in Web 2.0 is the development and evolution of Web-based communities and hosted services, such as social networking sites, wikis, Blogs and folkonomies. They can build on the interactive facilities of "Web 1.0" to provide "Network as platform" computing, allowing users to run software or applications entirely through a browser. Users can own the data on a Web 2.0 site and exercise control over that data.

The argument exists that "Web 2.0" does not represent a new version of the World Wide Web at all, but merely continues to use so-called "Web 1.0" technologies and concepts. Techniques such as AJAX do not replace underlying protocols like HTTP, but add an additional layer of abstraction on the top of them. Many of the ideas of Web 2.0 had already been featured in implementations on networked systems well before the term "Web 2.0" emerged:

1. Publish and Disseminate Information (Blogs, RSS, tag);
2. Network and Build Community (e.g. Social networking platforms like Flickr, Friendster, and MySpace);
3. Collaborate with Others (e.g. Exploring the World of Wikis);
4. Share Your Stories with the World (e.g. How to Record, Edit, and Promote a Podcast).

Nowadays the Web 2.0 covers extremely broad aspects in terms of theoretical, engineering and practical issues. Particularly, in this panel we will discuss what the new challenges in the Web 2.0 from the engineering point of view are.

The panellist will address a variety of issues incurred in Web engineering and applications such as:

- Whether modelling is still mandatory;
- How the development process must be re-shaped to accommodate to the new requirements and possibilities;

S. Hartmann et al. (Eds.): WISE 2008, LNCS 5176, pp. 183–184, 2008.

- How to efficiently support richer user interaction;
- What is about the web community discovery and analysis in Web 2.0;
- How to deal with the social network analysis in Web 2.0;
- Whats the trustworthiness evaluation in Web 2.0 information;
- How to improve Web search in the Web 2.0;
- Whats the current trend toward enabling unskilled users to develop own applications on the Web, how to efficiently enable them to do so and so on.

Author Index

Lecture Notes in Computer Science

Sublibrary 3: Information Systems and Application, incl. Internet/Web and HCI

For information about Vols. 1– 4801
please contact your bookseller or Springer

Vol. 4976: Y. Zhang, G. Yu, E. Bertino, G. Xu (Eds.), Progress in WWW Research and Development. XVIII, 699 pages. 2008.

Vol. 4969: R. Kronland-Martinet, S. Ystad, K. Jensen (Eds.), Computer Music Modeling and Retrieval. XII, 508 pages. 2008.

Vol. 4956: C. Macdonald, I. Ounis, V. Plachouras, I. Ruthven, R.W. White (Eds.), Advances in Information Retrieval. XXI, 719 pages. 2008.

Vol. 4952: C. Floerkemeier, M. Langheinrich, E. Fleisch, F. Mattern, S.E. Sarma (Eds.), The Internet of Things. XIII, 378 pages. 2008.

Vol. 4950: A. Kerren, J.T. Stasko, J.-D. Fekete, C. North (Eds.), Information Visualization. IX, 177 pages. 2008.

Vol. 4947: J.R. Haritsa, R. Kotagiri, V. Pudi (Eds.), Database Systems for Advanced Applications. XXII, 713 pages. 2008.

Vol. 4936: W. Aiello, A. Broder, J. Janssen, E.E. Milios (Eds.), Algorithms and Models for the Web-Graph. X, 167 pages. 2008.

Vol. 4932: S. Hartmann, G. Kern-Isberner (Eds.), Foundations of Information and Knowledge Systems. XII, 397 pages. 2008.

Vol. 4928: A.H.M. ter Hofstede, B. Benatallah, H.-Y. Paik (Eds.), Business Process Management Workshops. XIII, 518 pages. 2008.

Vol. 4918: N. Boujemaa, M. Detyniecki, A. Nürnberger (Eds.), Adaptive Multimedia Retrieval: Retrieval, User, and Semantics. XI, 265 pages. 2008.

Vol. 4903: S. Satoh, F. Nack, M. Etoh (Eds.), Advances in Multimedia Modeling. XIX, 510 pages. 2008.

Vol. 4900: S. Spaccapietra (Ed.), Journal on Data Semantics X. XIII, 265 pages. 2008.

Vol. 4892: A. Popescu-Belis, S. Renals, H. Bourlard (Eds.), Machine Learning for Multimodal Interaction. XI, 308 pages. 2008.

Vol. 4882: T. Janowski, H. Mohanty (Eds.), Distributed Computing and Internet Technology. XIII, 346 pages. 2007.

Vol. 4881: H. Yin, P. Tino, E. Corchado, W. Byrne, X. Yao (Eds.), Intelligent Data Engineering and Automated Learning - IDEAL 2007. XX, 1174 pages. 2007.

Vol. 4877: C. Thanos, F. Borri, L. Candela (Eds.), Digital Libraries: Research and Development. XII, 350 pages. 2007.

Vol. 4872: D. Mery, L. Rueda (Eds.), Advances in Image and Video Technology. XXI, 961 pages. 2007.

Vol. 4871: M. Cavazza, S. Donikian (Eds.), Virtual Storytelling. XIII, 219 pages. 2007.

Vol. 4868: C. Peter, R. Beale (Eds.), Affect and Emotion in Human-Computer Interaction. X, 241 pages. 2008.

Vol. 4858: X. Deng, F.C. Graham (Eds.), Internet and Network Economics. XVI, 598 pages. 2007.

Vol. 4857: J.M. Ware, G.E. Taylor (Eds.), Web and Wireless Geographical Information Systems. XI, 293 pages. 2007.

Vol. 4853: F. Fonseca, M.A. Rodríguez, S. Levashkin (Eds.), GeoSpatial Semantics. X, 289 pages. 2007.

Vol. 4836: H. Ichikawa, W.-D. Cho, I. Satoh, H.Y. Youn (Eds.), Ubiquitous Computing Systems. XIII, 307 pages. 2007.

Vol. 4832: M. Weske, M.-S. Hacid, C. Godart (Eds.), Web Information Systems Engineering – WISE 2007 Workshops. XV, 518 pages. 2007.

Vol. 4831: B. Benatallah, F. Casati, D. Georgakopoulos, C. Bartolini, W. Sadiq, C. Godart (Eds.), Web Information Systems Engineering – WISE 2007. XVI, 675 pages. 2007.

Vol. 4825: K. Aberer, K.-S. Choi, N. Noy, D. Allemang, K.-I. Lee, L. Nixon, J. Golbeck, P. Mika, D. Maynard, R. Mizoguchi, G. Schreiber, P. Cudré-Mauroux (Eds.), The Semantic Web. XXVII, 973 pages. 2007.

Vol. 4823: H. Leung, F. Li, R. Lau, Q. Li (Eds.), Advances in Web Based Learning – ICWL 2007. XIV, 654 pages. 2008.

Vol. 4822: D.H.-L. Goh, T.H. Cao, I.T. Sølvberg, E. Rasmussen (Eds.), Asian Digital Libraries. XVII, 519 pages. 2007.

Vol. 4820: T.G. Wyeld, S. Kenderdine, M. Docherty (Eds.), Virtual Systems and Multimedia. XII, 215 pages. 2008.

Vol. 4816: B. Falcidieno, M. Spagnuolo, Y. Avrithis, I. Kompatsiaris, P. Buitelaar (Eds.), Semantic Multimedia. XII, 306 pages. 2007.

Vol. 4813: I. Oakley, S.A. Brewster (Eds.), Haptic and Audio Interaction Design. XIV, 145 pages. 2007.

Vol. 4810: H.H.-S. Ip, O.C. Au, H. Leung, M.-T. Sun, W.-Y. Ma, S.-M. Hu (Eds.), Advances in Multimedia Information Processing – PCM 2007. XXI, 834 pages. 2007.

Vol. 4809: M.K. Denko, C.-s. Shih, K.-C. Li, S.-L. Tsao, Q.-A. Zeng, S.H. Park, Y.-B. Ko, S.-H. Hung, J.-H. Park (Eds.), Emerging Directions in Embedded and Ubiquitous Computing. XXXV, 823 pages. 2007.

Vol. 4808: T.-W. Kuo, E. Sha, M. Guo, L.T. Yang, Z. Shao (Eds.), Embedded and Ubiquitous Computing. XXI, 769 pages. 2007.

Vol. 4806: R. Meersman, Z. Tari, P. Herrero (Eds.), On the Move to Meaningful Internet Systems 2007: OTM 2007 Workshops, Part II. XXXIV, 611 pages. 2007.

Vol. 4805: R. Meersman, Z. Tari, P. Herrero (Eds.), On the Move to Meaningful Internet Systems 2007: OTM 2007 Workshops, Part I. XXXIV, 757 pages. 2007.

Vol. 4804: R. Meersman, Z. Tari (Eds.), On the Move to Meaningful Internet Systems 2007: CoopIS, DOA, ODBASE, GADA, and IS, Part II. XXIX, 683 pages. 2007.

Vol. 4803: R. Meersman, Z. Tari (Eds.), On the Move to Meaningful Internet Systems 2007: CoopIS, DOA, ODBASE, GADA, and IS, Part I. XXIX, 1173 pages. 2007.

Vol. 4802: J.-L. Hainaut, E.A. Rundensteiner, M. Kirchberg, M. Bertolotto, M. Brochhausen, Y.-P.P. Chen, S.S.-S. Cherfi, M. Doerr, H. Han, S. Hartmann, J. Parsons, G. Poels, C. Rolland, J. Trujillo, E. Yu, E. Zimányie (Eds.), Advances in Conceptual Modeling – Foundations and Applications. XIX, 420 pages. 2007.